SHELTER

Dear Mr. Wilson,
This book is a small token
of my appreciation for the
great job opportunity that I
received from you.
I would like to thank you
for allowing me to carry it
through three summers which
enriched my knowledge of architecture enormously.

Thank you
Vanessa Bigador.

Aug. 30/96.

SHELTER

Human habitats from around the World

CHARLES KNEVITT

POMEGRANATE ARTBOOKS
San Francisco

Acknowledgements

The author would like to thank the following for their contributions:
Clive Jackson and Hugh Bessant, Redland Roof Tiles, for commissioning the book; the production team for their dedication and good humour under pressure; Anne Cowlin, my personal assistant, and Wendy Nichols of PWP; Ceri Jones, Architecture and Planning Assistant, The Prince of Wales's Office; Sheila McKechnie, John Trampleasure and Victoria Bowman, of Shelter, the national campaign for homeless people; Dr Michael Roaf and Dr Susan Roaf; John F.C. Turner; Paul Oliver; Paul Finch, editor, *The Architects' Journal*; Professor Richard England and Mrs Myriam England; Mick and Elodie Waite; Mel Agace; Alexander and Mary Creswell; all the picture librarians who went out of their way in search of shelters; Roy A. Giles; Bernard Rudofsky and Myron Goldfinger whose books inspired this one; and my wife, Lesley.

PHOTOGRAPHS

The author and publishers thank the following for permission to reproduce the illustrations which appear on the pages listed.

Avedissian Akaa/Aga Khan Library, Geneva 135
© B. & C. Alexander 9 *(bottom)*, 49
Arcaid, London: Reiner Blunck 143; Richard Bryant 89, 93, 115, 137, 141, 147; Niall Clutton 91, 125; Colin Dixon 105; Richard Einzig 117; Scott Frances/ESTO 129; Farrell Grehan 51; Ben Johnson 81; Ken Kirkwood 109; Kurnewal/PRISMA 75; Ian Lambot 165; Lucinda Lambton 61, 113; Bill Maris/ESTO 103; Ezra Stoller/ESTO 107
Architectural Association, London: B. Chaitkin 9 *(top)*; Ron Herron 13 *(top)*; G. Smythe 101; Robert B. Vickery 99; Nathan Willock 97
© Steven Brooke 6, 139
Martin Charles 111
Charleston Postcard Company 123
Christie's, London/Bridgeman Art Library, London *frontispiece*
Peter Clayton 17 *(top left)*
Alexander Creswell 87
Richard Davies/Norman Foster & Partners 167
C. M. Dixon 19, 21, 33
Mary Evans Picture Library 15 *(top)*
Richard Foster/Trimedia Enterprises 65
Lásló Geleta 145
Roy A. Giles 59
Myron Goldfinger 41, 47, 53
Robert Harding Picture Library, London 1, 43, 57, 73, 77, 85, 119, 149, 151, 155, 157
Hutchison Library, London: 63; Tim Beddow 10 *(bottom)*; S. Errington 17 *(bottom)*; Goycoolea 13 *(bottom)*; John Hatt 29; J. Highet 15 *(bottom)*; J. Horner 14 *(bottom)*; C. Pemberton 16 *(bottom)*; Tony Souter 153

Simon Jauncey 95
A. F. Kersting 159
National Maritime Museum, London 83
National Trust Photographic Library/Mike Williams 10 *(top)*, 79
Martin Pawley 16 *(top)*
Photo Library International, Leeds 71
Armando Salas Portugal 131
Dr Susan Roaf 17 *(top right)*
The Royal Collection © Her Majesty The Queen 121
Royal Geographical Society, London: Anna Buckley 25; Chris Caldicott 45; David Constantine/Sitting Image 12 *(bottom)*, 31, 35; Drew Geldart 5; J. Holmes 3; Sassoon/Royal Geographical Society 27
Royal Ontario Museum, Toronto/Bridgeman Art Library, London 39
Shelter/Clive Farndon 8
Leigh Simpson/Porphyrios Associates 69
Space Frontiers/Telegraph Colour Library 12 *(top)*
François Spoerry Collection 133
Frank Spooner Pictures/Chip Hires 168
Still Pictures, London: © B. & C. Alexander 9 *(bottom)*; Mark Edwards 14 *(top)*; P. Hermanson/Samfoto 4; Stephen Pern 37
John F.C. Turner 67
Penny Tweedie/Telegraph Colour Library, London 23
Victoria & Albert Museum, London/Bridgeman Art Library, London 55
Vienna Slide/Harald A. Jahn 163

The illustrations on pages 11 and 127 have been provided by the author.

Due acknowledgment is made to Chambers Harrap Publishers Ltd for the use of the word *shelter* from Chambers Concise Dictionary.

Shelter: Human habitats from around the World
Production Team
Design Malcolm Preskett   Production John Taylor
Research Ann Taylor (Pictures) Elizabeth Turner (Buildings)
Printed in Korea

Uncaptioned Pictures

References are to pages

2. Detail of *Adoration of the Magi*; Pieter Balten (c. 1525–98)
3. Tent City for Nepalese Porters, Tibet
4. Lavvo Tent, Norway
5. Hanseatic Wharf, Bergen, Norway
6. Seaside Roofscape, Florida, USA
7. Miyoshi Farmhouse, Hokkaido, Japan
8. Unidentified homeless person, Lincoln's Inn Fields, London, England
9. *(top)* Villa Savoye, Poissy, France
168. Flood victims, Bangladesh

CONTENTS

Dedication 8

Preface 9

Index 168

SHELTER

Shelter – For those without

shelter *sheľtər*, *n.* a shielding or screening structure, esp. against weather: (a place of) refuge, retreat, or temporary lodging in distress: asylum: screening: protection.—*v.t.* to screen: to shield: to afford asylum or lodging to: to harbour.—*v.i.* to take shelter.—*adj.* **sheľtered** affording shelter.—*n.* **sheľterer.**—*n.* and *adj.* **sheľtering.**—*adj.* **sheľterless.**

PREFACE

SHELTER is something that we in the developed, Western world all take for granted – from the shelter of our mother's womb, through life, to the shelter of the grave. The house is our first experience of architecture, and by the age of six our rapidly growing awareness of the immediate environment in which we live is represented in simple drawings of our family, our pets, our garden and our home. In Christian households we soon learn about a family less fortunate than our own, seeking shelter under the most difficult circumstances and having to make do: 'And she brought forth her firstborn son, and wrapped him in swaddling clothes, and laid him in a manger; because there was no room for them in the inn'. As we grow up we then establish a home of our own, be it in a bedsit, student squat, local authority tower block or housing association flat, semi-detached suburban mock-Tudor 'Mon Repos', or detached five-bedroom des. res. with all mod. cons., bathrooms en-suite and double garage to boot. If we achieve even greater material success in later life, we might well be drawn 'back to the land' and wish to display our affluence by purchasing a country estate, where all we survey from our mansion windows is part of our territory, our personal Shangri-La. And towards the end of our days, through choice or out of necessity, we may end up in 'sheltered housing', preparing for the final journey to our eternal abode.

The history of shelter is the history of man, around the world and throughout time. At its most basic, the human habitat provides security and comfort. Civilisation has, however, imbued it with many other meanings to do with wealth and status, family life and notions of community and privacy, with individuality, identity and pride, but also with conflict, loneliness and despair. The house is an extension of our body, our personal space, a place of sanctuary

Cree Indian wigwam
Quebec
Canada

Whatever form it takes, shelter provides for our basic needs of security and comfort. A medieval cruck cottage at Quarry Bank Mill, Styal, Cheshire.

and our most precious psychological territory. For some people, it is also their most treasured possession; for others, merely a transit camp – a soulless shell – to be traded in almost as often as their car as they climb the property ladder; for others still, a source of worry and stress, both *in*secure and *un*comfortable. When it is a source of happiness and reassurance – when we occupy it and control it, and especially when we build it from scratch like many other animals – then we can do so with extraordinary resourcefulness, ingenuity and creativity, even playfulness. Great minds have dwelt on its significance, from Confucius and Aristotle to Leonardo da Vinci and Thomas More. Both the Bible and Shakespeare brim over with references. Even Machiavelli, whose domesticity has never really come to the fore, had this to say about retreating to his villa: 'When evening comes, I return home and go to my study. On the threshold I strip off my muddy, sweaty, workaday clothes and put on the robes of court and palace, and in this graver dress I enter the antique courts of the ancients and am welcomed by them'. The Florentine's home was, evidently, as much his castle as any Englishman's.

What of the dispossessed, the home*less?* If shelter is so deeply-rooted in our psyche and in our genes, what are the consequences for those without? For those of us old enough to remember it, the dramatised BBC television documentary *Cathy Come Home,* written by Jeremy

Troglodyte caves
Tunisia

Sandford and first broadcast in 1966, gave a first glimpse of how easy it can be to fall into that predicament. It was a harrowing account of a young woman who loses both her home and her family. Such was its impact that it was directly responsible for the creation of the charity for homeless people, Shelter. Subsequent research, referred to in psychiatrist Dr Anthony Fry's *Safe Space* (1987), has equated homelessness with increased rates of mental illness, alcoholism, epilepsy and tuberculosis. Clearly then, the loss of the security and comfort which shelter provides takes a heavy toll in physical and mental health. 'Yet the hardest of people in the toughest of cities still remain human', Fry wrote. 'We cannot fully deny that heritage. The soft under-belly still has to be protected or fed. The breast still needs an infant to suckle. The hardest man longs for the soft touch of another human being, and at night when all the screens stop flashing and the last video cassette has finally come to an end, the illusion of electronics and imagination must give way to aching limbs, tired eyes, weariness and sleep … There's no place like home!'

So, as Bernard Rudofsky wrote in *The Prodigious Builders* (1977): 'The house looms large, if not as a refuge, as a metaphor, live, dead and mixed. It is the repository of our wishes and dreams, memories and illusions. It is, or at least ought to be, instrumental in the transition from being to well-being.'

That evolution 'from being to well-being' has only been a conscious goal in the very recent past, the merest speck of time in relation to life on earth. To consider one's own lifespan in relation to all that has gone before, is a hum-bling experience. Planet earth has been in existence some 4,600 million years, and life some 2,000 million years. The first mammals date from 65 million years ago, the earliest hominids less than 2 million years ago, and *Homo sapiens* — fully modern man — only 100,000 years ago. Britain became an island, when its land connection with the Continent was severed by melting ice-sheets, only 7,000 years ago. Although there is evidence of modern man's predecessors marking out some type of territory for themselves, and devising the most basic shelters, almost a couple of million years ago, the real emergence of human settlements has only taken place in the last 10,000 years.

What brought that about was the transition from tribal hunting to farming and with it a more settled way of life. The land was managed, animals and plants became dom-esticated and, for the first time, the food surplus could be stored. 'In a mere 100 centuries,' wrote Desmond Morris in *The Human Animal* (1994), 'we progressed from mud huts to skyscrapers, in a breathless rush towards some imagined technological paradise.' More than that, of course, we began to conquer space – 'the final frontier' – and took the first steps towards colonising another 'planet', the moon.

Troglodyte cave house
near Saumur
France

II

PREFACE

From mud huts to moon shots in just 10,000 years. An artist's impression of an inflatable lunar habitat of the near future.

Imagine the 100 centuries to which Morris refers as a journey by air, say from Sydney, Australia, to London – a distance of a little more than 10,000 miles. Just after take-off, when the world's population was about 10 million and round mud huts covered in branches and skins start to become rectangular, you are crossing the Tropic of Capricorn near Alice Springs. Thereafter, doors and windows begin to be added to the walls and internal space begins to be sub-divided. Approaching the Indian Ocean, the 'room' becomes reality. This is the beginning of different uses for different spaces and therefore of privacy. Just past the equator, heading for the Horn of Africa, the Great Pyramids of Egypt are being built, soon to be followed by Stonehenge. Already more than three-quarters of the way through your journey, over Cairo, the Great Wall of China is being constructed by linking together lengths of existing walls.

Passing over Athens, with just 1,500 miles to go, the Roman Empire is drawing to a close; while over Rome itself, the great cathedrals of the Middle Ages are being built. By Florence, appropriately enough, the Renaissance is underway. At Paris, the Industrial Revolution is in full swing. The Modern Movement in architecture is just getting into its stride at Dover; and at Maidstone, 40 miles from touchdown, the end of the Second World War is heralding an unprecedented period of peace and prosperity in Western Europe, the consumer society and the post-war building boom. Man's 'three score years and ten', by comparison, is the distance between London and the orchards of deepest Kent, east of Ashford and south of Canterbury. Our concept of what shelter is – or 'ought to be', in Rudofsky's words – is clearly in its infancy.

In spite of 10,000 years of development and progress, the simple fact of the matter is that perhaps half the world's

12

Rural houses
Comilla
Bangladesh

population *still* lives in mud huts. There is a simple explanation for this: the population explosion. It took almost 12,000 years, from 10,000 BC to 1850, for the numbers of people to grow from around 10 million to 1,000 million. They doubled between 1850 and 1950; then doubled again between 1950 and 1981, just 31 years, to 4,000 million. Most recent estimates suggest that the population is increasing at the rate of 94 million a year, faster than ever, and will reach 8,500 million by 2025. As a result, Third and First Worlds share a common problem of sheltering ever-increasing numbers: overcrowding. It began with the rapid urbanisation brought on by the Industrial Revolution. In the first 50 years of the nineteenth century the population of London, then the capital of the world, doubled from one to two million. Other successful First World cities have followed suit so that now Los Angeles is, in area, half the size of Belgium. Cairo, with a population of 15 million, is expected to grow to 20 million within 10 years. But impressions can be deceptive. Surprising as it may seem, taking England and Wales together as one country, it becomes the third most densely populated country in the world after Bangladesh and Taiwan.

The implications of all this are dramatic and profound. Until quite recently, the First World response was to put its faith in technology as if it had all the answers to meeting population growth and material well-being. A high price

A technological fantasy of a terrestrial shelter, 1964. British architect Ron Herron's Walking City on the Ocean.

was paid, of course, not only in terms of finance and resources, but because in the enthusiasm for 'progress' other equally important issues came to be overlooked. Foremost among these was the need to satisfy man's primal concerns for territory, security and comfort by building, adapting and conditioning his environment to meet that need, rather than by trying to shoe-horn him into what was simplest, cheapest or most expedient. Better still, man

Nomadic tent
Rajasthan
India

World population growth is currently 94 million people a year and rising faster than ever before. 'Home' for this Calcutta family is a stack of drainpipes.

would have been offered the means to satisfy those basic needs himself rather than being forced to live with the consequences of other people's values and actions. The division of labour and increasing specialisation which the Industrial Revolution introduced, formed a wedge between man and his habitat. Third parties intervened between the need for shelter and the end product. An industry based on standardisation and prefabrication grew up, driven by means of production, access to raw materials and access to markets. Places began to lose their unique qualities so that everywhere began to look and feel like everywhere else. Culture became homogenised. Identity became subsumed, a sense of belonging was lost. Fashion and dogma became universal and unresponsive to local conditions; they over-whelmed local character. Settlements which had hitherto grown organically over time, became rigid grids of develop-ment – instant, impersonal, dehumanising and oppressive in scale and relentless in their brutality. Health and hygiene may have improved out of all recognition, but a higher price has had to be paid in other aspects of the provision of shelter. While individuals who commission their own home, or who can afford to live in a place of their choosing, can have virtually whatever they want, these luxuries are not available to a proportion of the world's population which is increasing exponentially.

A Luddite reaction against the trends of the past 200

Painted houses
Saqqara
Egypt

years would be impractical and ultimately counter-productive. Changes are manifestly needed, nonetheless, to restore the balance between the basic requirements of shelter, psychological as well as physical, and its provision. Remedies are easier to prescribe than to administer, but the first task is to disaggregate the problem and to tackle specific issues in specific locations. At the risk of over-simplification, this entails learning from the vernacular traditions of working with nature, climate and local materials, and from the 'genius of builders who never made architecture's *Who's Who*', as Rudofsky wrote. It means encouraging greater self-sufficiency and labour-intensive building techniques, through self-help and self-build; making shelters more responsive to local needs through participation and the community architecture approach; and recycling society's detritus into usable building materials. It also means reducing energy-consumption (and therefore pollution) in the manufacture and transportation of building materials, and in the buildings themselves; and ensuring that whatever is built is sustainable.

Most of these issues have already been championed more eloquently by others, not least by the Prince of Wales. In previous writings I have likened his role to that of St Ignatius Loyola, spearheading the Counter-Reformation and refusing to be intimidated by certain vested interests, whether in architecture, town planning and property

Health and hygiene preoccupied newly-urbanised Britain in the nineteenth century. Prince Albert built 'model' cottages for the Great Exhibition of 1851, with good sanitation, thermal insulation, proper ventilation and damp-proofing.

Gold-miner's shack
Rockies
USA

Recycling consumer society's detritus into usable building materials. A 'garbage' house built by British architect Martin Pawley and his students at Troy, New York, in 1976, from newsprint, bottles, steel cans and scrap neoprene rubber.

development; or, for that matter, in architectural criticism; critics have a tendency to hide their nakedness behind a Potemkin Village of print. Sooner or later the radical stance he has adopted in architecture, as in other fields, will become the new conventional wisdom and subsequent generations will wonder what all the fuss was about. The most reassuring remark by Le Corbusier, 'the architect of the century', was not about 'machines for living in' but that 'it is always life that is right and the architect who is wrong'. In future, less emphasis needs to be placed on the form and packaging of shelters, more on their content – and in the context of global needs.

In the brief survey of 75 shelters which follows, some of the extraordinary diversity of man's – and woman's – responses to the provision of places of security and comfort is explored. The examples range from the indigenous, unsophisticated and spontaneous to the universal, principled and highly contrived; from the most primitive and rudimentary caves and huts of Africa, Europe and Asia to the terrestrial inferno of Hong Kong's Walled City, and Millennium Tower, which could soon be the world's tallest skyscraper, half a mile high, at the northern boundary of the burgeoning Pacific Rim. Progression and regression are both illustrated. But as we take stock of our recent past, acknowledging the mistakes as well as the successes, and become more aware of such relatively new concerns as coping with population growth and overcrowding, countering unbridled pollution and making the planet more liveable, we should also celebrate our continuing ability to overcome adversity and create shelters which still have the power to hold us in awe and capture our imagination. 'The most powerful drive in the ascent of man is his pleasure in his own skill,' wrote Dr Jacob

Government flats
Bombay
India

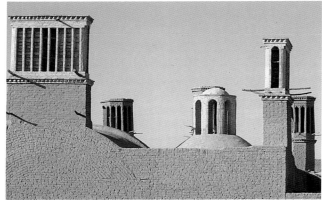

Learning from vernacular traditions. Windcatchers, called *badgirs*, are depicted on the roof of the House of Nakht in ancient Egypt *(left)*. These harnessed nature in an early version of air conditioning, 2,000 BC. *Right:* Twentieth-century *badgirs* carry on the tradition in Aliabad, Iran, from *The Windcatchers of Yazd* by Dr Susan Roaf, 1988.

Bronowski in *The Ascent of Man* (1973). 'He has to do what he does well and, having done it well, he loves to do it better. You see it in his science. You see it in the magnificence with which he carves and builds, the loving care, the gaiety, the effrontery. The monuments are supposed to commemorate kings and religions, heroes, dogmas, but in the end the man they commemorate is the builder.'

Having completed the first stage of our recent 'journey' of 10,000 years, from Sydney to London, we are now in a transit lounge awaiting the next, to – we know not where. In the meantime it would serve us well to reflect on the words of Henri de Saint-Simon, who wrote: 'The Golden Age of the human race is not behind us at all, it is ahead, it lies in the perfection of the social order; our forefathers have never seen it; our children will reach there some day. It is up to us to trace the path for them.'

Rural houses
Sichuan
China

Victoria Cave, Settle
North Yorkshire, England

*c.*10,000–8,000 BC

MAN's fascination with the most primitive forms of shelter is constantly being fed by new discoveries and by its depiction in popular entertainment such as the films *One Million Years BC* and *The Flintstones*. The oldest known inhabited cave in Europe is at Vallonet in southern France, first occupied between 900,000 and 950,000 years ago. West of Peking, China, is the Zhoukoudian cave measuring 175 metres (574 feet) long by 50 metres (164 feet) wide. This was first inhabited about 700,000 years ago and the remains of more than 40 examples of *homo erectus* were found there.

Many others, in the Middle and Far East, go back 60,000–250,000 years, and cave art – engravings and sculpture in bas-relief, as well as paintings – has been traced to 20,000 to 35,000 years ago, during the Ice Age. This cave, in the limestone hills two miles north-east of the Yorkshire town of Settle, was known to have been occupied by hunters about 10,000 BC, but is minute by comparison.

Nature was both hostile and provided some means of protection against the elements and predators. Caves were often used as temporary shelters and were easy to defend, especially if a fire was lit at the mouth. The control of fire dates back at least 460,000 years – possibly more than a million years.

But there is evidence to suggest that other basic shelters pre-dated even the use of caves. At Olduvai Gorge, in northern Tanzania, what appears to be the foundation ring of a circular hut is believed to be 1.8 million years old and various forms of windbreaks and skin tents were almost certainly lived in during summer months and warmer interglacial periods.

You are a king by your own fireside, as much as any monarch in his throne.

SAAVEDRA MIGUEL DE CERVANTES
Don Quixote (1606–15)

Mammoth-Bone Huts Mezhirich, Ukraine

*c.*16,000–10,000 BC

IN the treeless Arctic regions known as tundra, mammoths were useful for more than their meat and skins: their bones and tusks were often used as a basic building material, packed tightly together. The reconstruction of a hunter-gatherer's home at Mezhirich shows what a typical hut may have looked like. Further south and west, for example in the Grotte du Renne in northern France, tusks were found at encampments within caves – shelters within shelters – dating back 35,000 to 50,000 years. By about 10,000 BC most mammoths had become extinct, however, due to climatic changes and through being hunted. One carcass could feed several people for months.

Tusks formed the wall and roof arches and defined the entrance, with other bones being used as an infill before skins were added for weatherproofing. The overall diameter of the hut was probably four metres (13 feet) or more and excavations revealed a total of 385 bones around a central hearth, which had further upright bones, possibly for use in supporting cooking pots. Mezhirich is only one of many similar mammoth-bone huts discovered by archaeologists in the Ukraine.

Earlier still, and further west in France, stitched hides were draped over conical tents made of a framework of wooden poles, suggesting the technique of building tents travelled from west to east. Shortly after this period, in Siberia, bones from different types of animals were used as a framework for hanging skins.

It is the same as is the case with a house: the rationale will be something like 'A covering preventative of destruction by wind, rain and sun'. But while one philosopher will say that the house is composed of stones, bricks and beams, another will say that it is the form in these things for the given purposes.

ARISTOTLE *De Anima*

Bedouin and Berber Tents
Arabia and North Africa

Nomadic peoples require only the most basic of shelters, either adapting their immediate environment and then moving on, leaving their temporary home behind – for example, the Eskimos with their igloos; or making the best use of what they can carry with them – for example, the North American Indians with their tepees, and the Bedouin and Berbers of Arabia and North Africa with their tents, in which case they must be simple and quick to erect and dismantle.

Berbers are the aboriginal people of North Africa, largely scattered tribes of Morocco, Algeria, Tunisia, Libya and Egypt. As Eric Newby recorded in *On the Shores of the Mediterranean,* some are 'as black as Negroes from Central Africa. Some have blue eyes and rosy cheeks, which may be due to the familiarity of their women with Christian mercenaries and slaves.' Most were sedentary farmers until the Middle Ages, when invading Arabs wrecked their economy and they became nomadic.

Bedouin, on the other hand, are literally those who wander the desert, more than a third of them in Yemen and south-west Arabia. Arabs are essentially city dwellers, so the Bedouin account for only a small number. Both Berbers and Bedouin are herders, moving their camels or donkeys, sheep, goats and sometimes cattle from pasture to pasture.

As timber is scarce, a minimum of poles (usually four) form the basic tent structure and then woven cloth (of goat-hair in the case of the Berbers, camel-hair in Bedouin tents) is stretched over the top, and the guy ropes – fastened to webbing straps – are pegged into the sand. As many as 40 goat-skins may be required. In other parts, Ethiopia for example, a frame of poles is used to make a humped dome-like structure which is then covered in woven mats of grass or palm fronds.

The first night, then, I went to sleep on the sand, a thousand miles from any human habitation. I was more isolated than a shipwrecked sailor on a raft in the middle of the ocean.

ANTOINE DE SAINT-EXUPÉRY *The Little Prince* (1945)

Open-Sided Houses
Vanuata, South Pacific

MALO, one of the northern islands of the 13 which make up Vanuata (it means 'Our Land Forever'), has been occupied for some 3,000 years. The islands spread out over 400 miles from north to south, little more than dots among the estimated 25,000 which feature in the vast expanse of the Pacific Ocean. Captain James Cook called the group the New Hebrides when he charted them in 1774, a name they kept until Vanuata became an independent republic within the Commonwealth in 1980.

Because of the mild climate only the most basic structures are required and, like the basic Samoan house, dwellings tend to be large sheds comprising timber posts (often trunks of evergreen breadfruit trees) and pitched roofs of fibrous material, such as coconut palms, but with no walls. Mats are sometimes stretched between the posts and are used as backrests.

This makes Vanuata distinctive from many of the other Polynesian islands where elaborate and often highly decorated shelters include the loaf-shaped huts of New Ireland; the circular huts made from pig-proof yam trees in Papua-New Guinea; the thatched houses with soaring gables, built on stilts in the water, and great domed round houses of Western Samoa; and the tunnel-vaulted hangars like upturned boats in Tonga. Many different types were recorded in a book on Duperrey's *Voyage of the Coquille* in the early 1820s.

The peoples of the area are nothing if not class-conscious and *tahunga,* the Polynesian word for 'expert', is applied equally to any type of specialist from architect-carpenters to priests. Payment for building work is in food and invitations to banquets and other celebrations during construction, a ritual followed in the West at 'topping-out' ceremonies when the highest part of the building has been reached. As Rudyard Kipling wrote, in *A Truthful Song:* 'How very little, since things were made/ Things have altered in the building trade'.

Each of the islands of the Pacific Ocean is different from the others because the islands are so dispersed and so small. In a sense, it is only their isolation which has enabled them to survive as they are today. They are anachronistic, but a model, perhaps, of certain early human communities.

TERENCE DIXON and MARTIN LUCAS
The Human Race (1982)

Masai Manyatta
Kenya and Tanzania

WOMEN are the builders of the Masai, nomadic cattle herders whose huts are arranged in a circle around their livestock in a *manyatta*. A thicket of thorn branches around the perimeter keeps out unwelcome predators, especially hyenas. Their huts, which take up to a month to build, are either oblong or square, often up to six metres (20 feet) in length, and are made from a frame of sticks and saplings, closely woven, and then covered in leaves, mud and dung, which keep away the termites. They need constant maintenance and repair, especially in the rainy season although today plastic sheeting is often used to make them more watertight.

The houses are low, about shoulder-height, and built for security, with a cramped doorway and dog-leg bends before one reaches the hearth where a constant fire burns, increasing the intensity of the heat. Smoke is released through a small opening above. Flies are also constantly present, and days are usually spent outdoors under the acacia trees.

During migrations in the dry seasons, smaller temporary *kraals* are built by the same method, and the tribe lives almost entirely off the milk, blood and occasionally meat of its herds. When Ngai, 'husband of the moon and creator of all things', according to legend, went to live in the sky above Mount Kilimanjaro, he left to the Masai all the cattle of the world for their succour. Rustling their neighbours' herds was therefore considered no more than taking back what was once theirs in any case. Cattle are also used as a dowry – to the wife, not her family – and husbands can have many wives.

On marriage, the new wife builds a house for herself with the help of her women friends in the village.

26

IN THE BEGINNING

The *manyatta* was a large one, composed, like all others, of longish low huts rounded like loaves. Made from a mixture of mud and dung plastered on a frame of curved sticks, they reminded me of dried-out chrysalises.

KUKI GALLMANN *African Nights* (1994)

Beehive Houses
Aleppo, Syria

*c.*6,000 BC

BEEHIVE houses are far from unique to the limestone hills around Aleppo, near Syria's border with Turkey. Variations can be found in different parts of the Mediterranean, in west and southern Africa, Lewis in the Outer Hebrides and in the broch of Mousa in the Shetland Islands, off the northernmost coast of Scotland. These prehistoric circular stone towers were built by tightly wedging stones together over a mound of earth, which was then scooped out and the floor level lowered to increase the headroom.

In northern Syria the form dates back 8,000 years and the houses are still lived in – and built – today. Unusually, during the Halaf period (*c.*6,000–5,400 BC) rectangular houses with many rooms were abandoned in favour of a return to round huts called *tholoi,* which sometimes had a rectangular entrance chamber or storeroom, and were keyhole-shape on plan. Walls divided the domed space into smaller rooms. Commonly they were three to seven metres (10 to 23 feet) in diameter, of stone or mud-brick and then coated in a mud and straw plaster. They kept their inhabitants warm in winter, when snow lay thick on the ground, and cool in the extreme temperatures of summer. Slit openings in the wall provided some level of interior daylight.

In his book *The Prodigious Builders,* Bernard Rudofsky quotes a visitor to Lewis, in 1880, finding himself among 'stone age' beehive houses grouped together like 'a Hottentot village rather than a hamlet in the British Isles'. Smoke escaped through the turf roofs rather than chimneys, and the soot that accumulated was scraped up and used as valuable manure.

> [The vernacular is] a communal art, not produced by a few intelligences or specialists, but by the spontaneous and continuing activity of a whole people with a common heritage, acting under a community of experience.
>
> PIETRO BELLUSCHI

Houseboats
predominantly South-East Asia

*c.*2,000 BC

WHEN Cleopatra – dressed as the goddess of love – called on Mark Antony at Tarsus (now in modern Turkey) she arrived by *thalamegus,* a luxurious barge kitted out like a floating palace for her and her retinue, and a type of houseboat common on the Nile for at least 4,000 years. But for literally millions of people, houseboats are still permanent dwellings and the floating cities of Hong Kong, Singapore and Vietnam (pictured here) are anything but romantic.

Where residential land is scarce in hot climates, the water provides a natural alternative, as a source of food, for washing, as a coolant and as an unmetalled highway. Many boat-dwellers are permanent refugees or the victims of discrimination, from the Tanka and Hoklo peoples who live in the Hong Kong districts of Aberdeen, Yaumatei and its typhoon shelters, to the British who originally settled in boats in the Vale of Kashmir because, as foreigners, they were forbidden to build houses on land.

Some 35,000 boat-dwellers in Hong Kong still eke out an existence on water, the more enterprising doing a lucrative trade smuggling consumer goods to mainland China, and gold and silver coins, herbal medicines and sometimes human cargoes back again. Here, and in Canton, the population is served by floating kitchens, floating barbers, floating doctors and even floating undertakers.

What is not generally appreciated is that many others in the Western world still live in houseboats, whether through choice or out of necessity. Early American settlers travelled the Ohio and Mississippi rivers with their families and livestock until they found land they wanted to settle, then used the boat's timber to build a house or to sell as fuel – hence the expression 'sold down the river'. Shanty-boats and floatshacks were common during the Great Depression of the 1930s, and only 25 years ago it was estimated that there were 5,000 people living on houseboats in Seattle, Washington State.

When I was at home I was in a better place; but travellers must be content.

WILLIAM SHAKESPEARE *Antony and Cleopatra*

Dry-Stone Houses
Skara Brae, Scotland

*c.*1,500–2,000 BC

IT took two great storms, in 1850 and 1926, to expose the well-preserved remains of a group of seven houses at Skara Brae on West Mainland, Orkney Islands, dating from the late Neolithic period. Undressed slabs of local stone, which splits naturally along its length, were assembled without mortar to form the rectangular buildings with rounded corners, to a height of about two-and-a-half metres (eight feet). Similar dwellings have been discovered in France from about the same period.

The stone was used not only for the structure, which was buried in man-made mounds of midden up to roof level, but for furniture as well: box beds, tanks (possibly for holding shellfish), and dressers. There were no trees on the island. The houses were connected by narrow passages which were also roofed in stone slabs and comprise one room each with a long narrow doorway which was probably closed off with another stone slab. A primitive sewage system connected up the drains from each house.

The foundations of even older huts were found beneath the walls, together with pottery decorated with a spiral, a common prehistoric motif found in other countries but this is the only example found in Britain.

Rectangular buildings can be traced back to 10,000 years ago and the earliest complexes of human settlements have been identified at Catal Hüyük, Turkey, in about 8,300 BC, and Jericho, in the Jordan Valley, in about 6,000 BC. But this discovery in the Orkneys, the Orcades of classical literature, provided important evidence about the period, along with standing stones, circles and earth houses also found locally.

Did not those night-hung houses,
    Of quiet, starlit stone,
Breathe not a whisper – 'Stay,
    Thou unhappy one;
Whither so secret away?'

WALTER DE LA MARE *The Suicide*

Tufa-Cone Houses
Urgup, Cappadocia, Turkey

IST CENTURY

WHILE the earliest records of Cappadocia date back to about 600 BC, the extraordinary volcanic rock formations which are still inhabited today probably date from about 2,000 years ago. Around AD 700 some 30,000 anchorites lived in the thousands of chambers carved out of the soft *tufa,* which is easily worked but hardens on contact with the air. What the ravages of the weather have started, man completes and continues to adapt.

One city used to burrow 80 metres (265 feet) underground. Another was the height of a 16-storey building. A six-mile tunnel connected two communities, one containing 20,000 people.

As well as fashioning houses, they sculpted stables and granaries, churches and monasteries of Byzantine exuberance, carved niches, applied Greek inscriptions and used the exterior walls as their canvas for paintings. Rock was hollowed out to make pigeon-cotes, then partly walled up again until removed so that the guano could be collected as a fertiliser. Building was a continuous process; new rooms and staircases were hollowed out with the arched blade of an adze. Flat-roofed cubic houses made from the same rock nestle among the conical towers, where access used to be by rope ladder.

The climate is often cold and inhospitable in this semi-desert region where the daily comforts we take for granted, such as piped water and central heating, have simply never existed.

O ye that dwell in Moab leave the cities, and dwell in the rock like the Dove that maketh her nest in the sides of the hole's mouth.

JEREMIAH chapter48, verse 28

Yurts
Central Asia

13TH CENTURY

SEEN from afar yurts, or *yurta,* have a solid appearance, but these circular tents of the nomadic Kazakhs of central Asia are portable and light. They can be set up or dismantled by a few men – or often women – in 30 minutes. When folded, they are carried on horseback or on a small wagon. The earliest known examples date from the thirteenth century.

The sloping walls are made up from a series of collapsible trellises just over one metre (four feet) high, which when expanded are secured by leather thongs. Space is left for a full-height doorway. A compression band of several ropes made of yak hair is then tied around the eaves to resist the outward thrust of the wooden poles which make up the truncated cone roof. Another compression band at the summit forms a central skylight-cum-chimney opening, although it is more common today to see a stovepipe rising through it.

The whole structure is then covered in skin, brightly-coloured handwoven textiles or felt, which is made by the women and has the properties of being lightweight, waterproof and wind-resistant. Finally, ropes are criss-crossed around the exterior of the whole tent to anchor it to the ground.

Internally, the one large space remains undivided except for invisible demarcation lines of etiquette. Writing in *Country Life* in 1973, Elisabeth Beazley described how the yurt is always pitched facing south, so that sunlight from the smoke hole in the roof acts as a sundial. The area between the hearth and the door is the entrance hall and the rest is divided into thirds. To the west is the women's part, to the east is the guest area, and at the back a mat is spread out for the men.

Now the yurts tend to be poor men's dwellings associated with villages, or they act as workshops for the richer, settled tribesmen. Less than 10 years ago it was a common enough sight to see groups of tents on the vast plain of the Gorgan; now it is a rare event.

ELISABETH BEAZLEY (1973)

Tepee, Great Plains of Canada and the USA

16TH CENTURY

NORTH American Indians, for the most part, lived in gabled houses of timber, earthlodges or the tiered adobe villages called *pueblos*. Hollywood is responsible for making us think otherwise. Wigwams, usually dome-shaped and covered in bark, were more permanent shelters than tepees, used by nomadic Indians hunting bison. The earliest recorded tepees date from the mid-sixteenth century, but by around 1700 they were more widespread, first as summer homes and then year-round.

These conical tents were a triumph of portability. Three, four or more poles, often pine, were easily carried or dragged by a horse, which also carried the bison hides, cloth mats or canvas used to cloak them. Bark was also used. At a new camp site the chief would determine the sacred geometry of the place, and any omens that were present, before the poles were erected in a matter of minutes. It was the women's responsibility to cover them as skins were a gift from the bride's family on marriage.

Sometimes there was an interior lining, providing some degree of thermal insulation and a secondary barrier against rain. Flaps at the top allowed smoke out and ventilation in without creating downdraughts, and in the worst weather a storm hood was placed over the apex. At the foot of the tent, pegs or stones would hold down the covering. Tepees typically ranged in height from three-and-a-half to six metres (12 to 20 feet).

Colourful scenes of animals, hunting and the weather (for example, rainbows and lightning) were painted onto the tepees during the non-hunting season in winter.

This detail, from Paul Kane's 'Indian Encampment on Lake Huron', shows tepees covered in bark.

You are the buffalo-ghost, the bronco ghost
With dollar-silver in your saddle-horn
The cowboys riding in from Painted Post,
The Indian arrow in the Indian corn ...

STEPHEN VINCENT BENET *John Brown's Body*

Cave Houses
Guadix, Spain

18TH CENTURY

GYPSIES have lived here in the caves at the foothills of the Sierra Nevada of southern Spain since the eighteenth century. Guadix, which was known to the Romans as Acci, means 'River of Life', and life must have gone on virtually unchanged for the tribe of 10,000 or so since that time.

There are similarities with the cave dwellings of Cappadocia, Turkey: the soft *tufa* hillocks have been eroded by time and climate into bizarre shapes, and the inhabitants have taken advantage of their surroundings to carve out small chambers which interconnect with one another. Some spread out through 20 'rooms', while others are combined with extensions of whitewashed walls and red-tile roofs. The whitewashed chimneys and exposed entrances are the only features which interrupt the landscape of soft limestone, clay and loam.

Troglodytism is still found in several other parts of the world, for example along the south bank of the Loire valley near Saumur, where in the last century half the population lived in their man-made caves; in Sicily, where cliff-face burial grounds made 3,000 years ago were later converted for the living; in Ethiopia and Egypt; and in several parts of China, where square holes up to four storeys deep are carved into the fields so that people live under their crops.

In very recent times, some 50 underground – or 'earth-sheltered' – buildings for various uses have been created in Britain, as many as 6,000 in North America. Reading about them brings back childhood memories of J.R.R. Tolkein's Hobbit, and Mole and Badger in Kenneth Grahame's *The Wind in the Willows,* in which the seasons come alive: 'Spring was moving in the air above and in the earth below and around him, penetrating even his dark and lowly little house with its spirit of divine discontent and longing'.

In a hole in the ground there lived a hobbit. Not a nasty, dirty, wet hole, filled with the ends of worms and an oozy smell, nor yet a dry, bare, sandy hole with nothing in it to sit down on or to eat: it was a hobbit-hole, and that means comfort.

J R R TOLKEIN *The Hobbit* (1937)

Reed Houses of the Marsh Arabs, Iraq

*c.*5,000 BC

THE Ma'dan, or Marsh Arabs, of southern Iraq occupy the river delta of the Tigris and Euphrates, successors to the Sumerians whose reed houses of 7,000 years ago barely differ from those still being built today. But for political reasons both the people and their unique form of untutored architecture are threatened species. Thirty years ago the explorer, writer and photographer, Wilfred Thesiger, predicted as much, as far as the houses were concerned, adding the important observation that 'long familiarity with houses such as these may well have given man the idea of imitating their arched form in mud bricks, as the Greeks later perpetuated wooden techniques in stone'.

In this region reed can grow more than six metres (20 feet) high and papyrus three metres (10 feet). The reed is cultivated to make matting (seen here in the foreground) as well as the houses, which sit on man-made islands of a mixture of earth and papyrus pressed hard together.

Construction of the houses is a simple process: thick bundles of reeds, called *fasces,* are bound together and embedded in the earth in two parallel rows as columns. The tops are then bent over and tied by a man using a tripod of reeds as a step-ladder, to create a parabolic arched tunnel. Lighter bundles are then lashed horizontally in between the columns; matting of split reed is added, and the two ends of the tunnel also filled in by the same method. Further columns are added at one end to denote the doorway. In his *Cultural Atlas of Mesopotamia and the Ancient Near East,* Dr Michael Roaf illustrates a drawing of the Late Uruk period, 5,000 years ago, proving the longevity of this technique.

Reed is a good insulant but has a comparatively short life. It is also flammable and insects love to use it as a nest.

> Sitting in the Euphrates *mudhifs,* I always had the impression of being inside a Romanesque or Gothic cathedral, an illusion enhanced by the ribbed roof and the traceried windows at either end, through which bright shafts of light came to penetrate the gloom of the interior.
>
> WILFRED THESIGER *The Marsh Arabs* (1964)

Dogon Villages Timbuktu, Mali

*c.*10TH CENTURY

SIGMUND Freud is unlikely to have been familiar with the Dogon, a tribe that has lived south of Timbuktu, bordering the Sahara desert, for between 600 and 1,000 years. For him *any* house was 'a substitute for the mother's womb, the first lodging, for which in all likelihood man longs, and in which he was safe and felt at ease'.

Flat-roofed houses, made of beaten clay and with timber columns and beams, sit four-square alongside circular granaries with conical thatch roofs made of millet straw. Like their dramatic sculpture and masks, the Dogon's buildings possess strong magical powers and are laid out in a certain configuration: villages are in pairs, representing heaven and earth, amidst fields of onions and millet laid out spirally in the form of a net. As far as the topography allows, the courtyard compounds follow an ideal pattern in which the smithy and men's meeting house, to the north, represent the head; the family houses represent the chest and belly; and altars, to the south, the feet. In between, the communal altar is a phallic symbol and the stone, used for crushing fresh ears of corn to make oil, represents female genitalia. Freud would have relished it all, but there's more.

The ground floor of each house is a woman lying on her back, ready for intercourse with a man, the ceiling. The façade of the house or granary is literally a face, often with an elaborately carved door; and the men's meeting house, or speech house, which fulfils the communal function of a village pub, often features timber columns carved in the form of well-endowed women.

Significantly for those of us preoccupied with houses as property and a hedge against inflation, a Dogon 'house' is not a structure but refers to the occupants, as far as the tribe is concerned. And such is its symbolism, a Dogon's dwelling-place is never sold.

And now here is my secret, a very simple secret: It is only with the heart that one can see rightly; what is essential is invisible to the eye.

ANTOINE DE SAINT-EXUPÉRY *The Little Prince* (1945)

Barrel-Vaulted Houses
Thíra, Greece

*c.*2,000 BC

THE molten mass of houses that flow like lava down the steep slopes of ancient Thera are an appropriate metaphor for the place itself – the remaining portion of a volcano. First occupied before 2,000 BC, the southernmost island of the Cyclades group was known in antiquity as Calliste, Most Beautiful; today it is also known as Santorin, Santorini and, in modern Greek, Thíra. For some, it is the lost city of Atlantis. The most recent eruption was in 1956.

As an example of the whitewashed architecture of the Mediterranean we have come to know and love, it has few equals apart from Mykonos and Amorgos. Stepped ramps, walls and small terraces snake their way between the barrel-vaulted houses, some semi-troglodytic, each one different from its neighbour, with timber doors and windows at every turn. The truly organic environment is peppered with rounded chimneys and sculpted church fronts and bell-towers.

Deep in plan, the floors, walls and ceilings of each house are also whitewashed to reflect light within it from the openings, usually a central doorway with a single window on either side and one above it. Blocks of lava were used to construct the walls; the roofs were made of local pumice stone (along with the local wines, made from vines which thrive in the volcanic ash, a principal export) and a mixture of the ash and lime as cement.

Visitors in the latter half of the last century, keen to examine the legacy of the island's explosive past, were struck by is 'primitivity'. But more perceptive observers had already noted that 'one ought to speak of *perfection* as expressed in a marvellous unity of past and present, which imparts to the stranger a sensation of timelessness'.

Between 180 and 400 metres (600 and 1,300 feet) below the uppermost parts, the Aegean Sea spreads out from a vast bay where three volcanic islands sit brooding, a reminder of its past – and possible future.

One cannot overemphasize the fact that everything – meaning and value as well as appropriateness of individual human conduct or the energy state of an atom – depends upon the interaction of the thing itself and its environment.

CYRIL STANLEY SMITH

Inuit Igloo
Canada and Greenland

For most of the year Eskimos *(Inuit)* live in cloth or sealskin tents and an igloo (also called an *aputiak*) is only built as a temporary home when they are hunting. It is an effective device for keeping out the elements, entered by a tunnel built below the floor level of the living space, so that cold air is carried away by convection. Snow, cut with a special knife, is the only readily available building material and, when formed into a dome, provides the largest volume for the smallest surface area, thus minimising heat loss through wind chill.

The diameter is usually two to five metres (six to 15 feet) and it is built by one or two people, sometimes a whole family if they are travelling together, in a matter of an hour or two. After cutting the blocks, measuring just over one metre by half that (four by two feet), the first layer is laid in a ring and the top planed to allow a spiral of further blocks to be added until the summit of the dome is reached, and a hole is left at the top for ventilation. Sometimes a window is inserted in the wall – a seal's intestine or a clear block of ice.

If the igloo is to be used for more than one night then its diameter can increase; for more permanent occupation, further domes of usually similar size are built alongside and connected by tunnels, but each one is only occupied by one person or one family. We tend to think of igloos as being made of snow, but the term (from the Eskimo *igdlu*, meaning house) is also used historically for domes in many different materials such as stone, sods of earth and timber.

Form is a mystery which eludes definition but makes man feel good in a way quite unlike social aid.

ALVAR AALTO Lecture to the Architects' Association of Vienna (1955)

Adobe Houses
New Mexico, USA

MUD, glorious mud, has been the most common building material in many parts of the world for thousands of years, as sod, cob, wattle and daub, pressed or rammed earth, and as bricks, sun-dried or burnt. It is abundant, easily worked and has excellent thermal insulation properties. Straw, or other fibrous material, is often added to help prevent shrinkage cracks as it dries out. Rain is its worst enemy, so its use is usually confined to regions with less than 25 centimetres (10 inches) a year, and often where timber is scarce. Mud houses are found around the Mediterranean – in southern Spain, North Africa and the Middle East – and in the Americas date from pre-Columbian times. They are found as far south as Peru.

Adobe refers to both the type of soil used, a heavy clay, and to the unburnt bricks themselves. When used as bricks the mortar is of the same material and the house is finished off in a coat of adobe, which dries to a hard, uniform mass, or lime or cement plaster. Walls are thick, as mud is structurally weak, and wall heights tend to be low. Pine poles, called *vigas,* are used as gently sloping rafters to hold up the roof, and these often project from the face of the building. Stone, or more recently concrete, provides a base to prevent capillary action.

Since around 1300 the Pueblo Indians of Central and South America have built settlements around a central courtyard using large adobe blocks. *Pueblo* means village or town in Spanish, and some examples are up to five storeys high. Rooftop terraces are formed by setting back each level from the one below, and access is by wooden ladders which project through a hole in the ceiling. Ground floors are traditionally for storage, particularly of grain.

The twentieth century has seen a revival of interest in adobe building, and the gentrification and modernisation of earlier examples, sometimes with garish results.

Inasmuch as there is nothing in bare earth to sell, no commercial group can be found to extol its merits.

KEN KERN

Trulli of Alberobello
Apulia, Italy

THE archaeologist Lancelot G. Bark, on visiting the beehive dwellings of Apulia in 1932, wrote: 'Stone in embarrassing profusion'. The *trulli* appeared 'unreal and wildly fantastic'. The latter, if not the former, observation, was undoubtedly true, even though similar structures have survived virtually unchanged for 4,000 years.

The megalithic houses originally had square or rectangular rooms within circular exterior walls, but these too eventually succumbed to the right-angle. The conical dry-stone roofs, in layered mounds, terminate in chalk-coated finials or cupolas alongside small towers of whitewashed chimneys which vie for attention. Older walls measure more than three metres (10 feet) thick, the more recent half that. Certainly they were a sturdy defence against climate enjoyed by rural peasants, although today many have been adopted by wealthier urbanites who crave ethnicity in their second homes.

Some *trulli* are whitewashed from head to toe; others have just painted walls or are left entirely natural. But the roofs, which swirl and collide in a slightly topsy-turvy fashion due to irregularities in the size and shape of the stones themselves, are usually made up of several cones, as many as one per room. Inside, the domed roof space is used for storage and there is a fresh-water cistern in the basement below the main living space.

Trulli are scattered across the landscape of vineyards, almond and olive groves, generally in units of three to 20; but in the Rione Monte district of Alberobello itself they snuggle up to each other to define the winding streets which radiate from the central square.

One may regard a work of architecture as a living organism in which all parts of its aspect ... must follow the same rhythm, for lacking this it runs a great risk of not being able to thrive.

ANDRÉ LURÇAT *Architecture*

Screen Houses
Japan

THE traditional Japanese house has changed little in more than a thousand years and this detail from a series of drawings illustrating the life and pastimes of the Japanese Court (Tosa School, c.1800) shows many of its key elements.

Simple post-and-beam construction lends itself well to rectilinear architecture which is both elegant and highly adaptive. As the walls are non-loadbearing they can easily be moved – or removed – to form open-plan spaces and allow the free-flow of air for ventilation. On a more spiritual level, the boundary between indoors and outdoors – between the man-made world and the natural world – evaporates as an enclosed and carefully defined interior becomes an open pavilion at one with its surroundings. Nature is respected and sometimes feared in an area prone to natural disasters such as earthquakes and volcanic eruptions.

The screens take several forms and are made of varying materials, left in their natural state: bamboo, rice or mulberry paper, cedar wood or clay plaster. They fold or are free-standing, slide along grooves or can be removed altogether. In the hottest climates, latticework or carved wooden panels are used. The low spreading eaves of the roof reflect the outstretched branches of a tree: Buddha sought enlightenment while resting under a tree.

And there is little concept of privacy in Japanese houses. Life is transitory so they are considered as mere temporary shelters.

As the Orient opened up at the turn of the century, its design traditions became a significant influence on contemporary architects – on Mackintosh, in the interiors of Glasgow School of Art; on the Prairie houses of Frank Lloyd Wright; on Gropius' ideas on dimensional coordination for curtain walling and modular design (based on the dimensions of the Japanese *tatami* mat); and generally because the qualities it represented lent themselves well to the industrialisation of the building process and the creation of pure forms purged of all decoration.

A CONTINUING TRADITION

While Western civilisation with its enormous technical achievements in building long ago succeeded in making life within the house independent of climate changes, in the Buddhist world nature has never been considered as something to be fought against, conquered, and mastered.

HEINRICH ENGEL *The Japanese House* (1964)

The Painted Ladies
San Francisco, California, USA

LATE-19TH TO EARLY-20TH CENTURY

IN LESS than 20 years the Victorian timber houses of San Francisco have achieved celebrity status thanks to what is known as The Colorist Movement, a quiet revolution in which 'the weapons are paint brushes and the ammunition is paint and imagination'. The results are The Painted Ladies – a term which was only coined in 1978, as the title of a book, but which is now used generically to describe any polychrome Victorian houses in timber found in 46 states across the USA.

Typically the terrace (or row) houses date from the middle of the last century and were built on standard lots of seven-and-a-half metres (25 feet) by 30 metres (100 feet). The designs and specifications came from pattern books published by architects, but surprisingly no two houses are identical. They are 'stick-built' in a 'balloon frame', held together by nails and using mass-produced elements from sawmills. Redwood or Douglas fir is soft and easy to cut but also very strong and the use of cut, then wire nails (typically 65,000 for a five-bedroom house), instead of mortice-and-tenon joints, could achieve cost savings of 40 per cent and also enable anyone to put up their own house rather than use skilled carpenters.

The burgeoning population of San Francisco, which quadrupled to almost a quarter of a million between 1860 and 1880, took readily to the simplified form of construction and some 50,000 balloon-frame houses were built. Of these, only 13,500 remain, the rest being the victims of the 1906 earthquake or mass clearance programmes of the 1950s and 1960s. But during the 1960s, the hippies began to apply amazing technicolour dreamcoats of paint to what were rather staid and subdued exteriors finished in cheap white Colonial Revival or Navy-surplus battleship grey.

Although the experimentation has continued, for example using gold leaf, aluminium and silver, today more subtle colours predominate.

A CONTINUING TRADITION

God took the beauty of the Bay of Naples, the Valley of the Nile, the Swiss Alps, the Hudson River Valley, rolled them into one and made San Francisco Bay.

FIORELLO LA GUARDIA Mayor of New York City

Painted Houses of the South Ndebele, South Africa

20TH CENTURY

P AINTING shelters is a common activity among many cultures, from Fiji and Papua-New Guinea to Egypt and parts of Yemen, and the tepees of the Blackfoot Indians of Montana. Often it is considered women's work, and many interiors are as brightly coloured and beautifully decorated as the exteriors. But the painted houses of the Ndebele are a comparatively recent phenomenon, dating only from the turn of the century.

No-one is quite sure where the tribe originated or when it moved to this part of Africa, between the Limpopo and Vaal rivers. What is known is that their original beehive-shaped structures were round and made from saplings covered with grass thatch, a type common in Africa since AD 800. Entrances were low and tunnel-shaped so that one had to stoop to get in. The dry-stone walled cattle byre was at the centre of the compound, emphasising the importance of the people's livelihood. More recently, the houses have been built in mud, become rectangular, the entrances have been enlarged and they are often celebrated with stepped or rounded architraves.

Ndebele have a great flair for design, for art and craftsmanship. The women decorate themselves with beaded hoops and brass rings, they wear patterned clothes and headdresses, and they produce the most wonderful murals on their homes. Sometimes the patterns are abstract, sometimes they depict human or animal forms, sometimes contemporary technology such as safety razor blades, street lamps, alarm clocks and motor-car number plates.

Originally subdued whites, yellows and dark greys were made from coloured clays, charcoal and cow-dung, and applied with feathers for paint-brushes. Later blue was added – from blue washing powder. Today the tribe tends to favour brighter and more durable acrylics, uses modern brushes and, with skill and practice, can paint a straight line up to 12 metres (39 feet) long freehand.

You white people are funny. You pay lots of money for a little piece of canvas, you paint on it, then hide it inside your house where no-one can see it! We Ndebele teach our daughters to paint for everyone to see ... and enjoy!

NDEBELE WOMAN quoted by Aubrey Elliott (1989)

Deanery Garden
Sonning, Berkshire, England

SIR EDWIN LUTYENS 1899–1902

BUILT for Edward Hudson ('dear Huddy'), the founder of *Country Life*, this house is in the continuing Arts and Crafts tradition of English architecture begun by Philip Webb with his Red House for William Morris. Significantly, Lutyens had recently completed a house at Munstead Wood, Surrey, for Gertrude Jekyll, his cousin, who revived the English cottage garden.

Their collaboration at Deanery Garden (now The Deanery) resulted in a marriage of country house and its setting 'at once formal and irregular', Christopher Hussey, Lutyens' biographer, wrote. 'Miss Jekyll's naturalistic planting wedded Lutyens' geometry in a balanced union of both principles.' It was, he said, 'a perfect architectural sonnet'.

Almost Tudor in design, the house is both picturesque and rooted in the vernacular. Contemporary craftsmanship was given full reign in the way oak was used extensively for the frame and great bay window of the two-storey hall, its flooring, and in the use of local tiles and red bricks, in this case from the Collier works in nearby Reading.

Lutyens was only 30 years of age at the time he received this commission, yet it demonstrates his supreme self-confidence and maturity as a designer. Facing the garden to the south-west is a deep arched entrance, approached by a bridge over a half-hidden circular pond with, on one side the bay window with its 48 leaded lights, and on the other a tall triple chimney stack. The interiors are equally compelling and a foretaste of many of his later country houses.

'So naturally has the house been planned that it seems to have grown out of the landscape rather than to have been fitted into it,' Hudson wrote the year after its completion. The Lutyens-Jekyll partnership continued and gave rise to a school of design, known as the Surrey style, in which formal hard landscaping soon became mellowed by the informal planting.

I went into a house, and it wasn't a house,
   It has big steps and a great big hall;
But it hasn't got a garden,
   A garden,
   A garden,
It isn't like a house at all.

A A MILNE *When We Were Very Young*

Commune Domes
South-Western States, USA

*c.*1965

OMES are among the most common form of shelter, almost as common as the circular plans of wigwams and igloos, African thatched huts and the original *trulli* stone houses of southern Italy. They are imbued with religious significance (denoting heaven, the sky or the universe) and later became symbols of secular and civic authority. By a certain irony, they also proliferated amongst the hippies and drop-outs of the burgeoning counter-culture of the United States in the late 1960s, especially in Arizona, Colorado, New Mexico and California.

'To live in a dome is – psychologically – to be in closer harmony with natural structure … Corners constrict the mind … Domes carry the values of the community into the outer world', members of Drop City, Colorado, told *Radical Technology* in 1976. Drop City was started by three former students from Kansas who bought some goat pasture and started to build domes out of scrap timber, car bodywork panels and windshields, and waterproofed tar paper. 'We find what we can use and use what we can find' was their motto. Most of these cheap and cheerful – and lightweight – DIY structures were geodesic, that is, made up of short, straight rods, or hexagonal panels, which when assembled in a certain configuration produce a curved surface or dome.

The first geodesic dome was designed by Dr Walter Bauersfeld for the planetarium on the roof of the Carl Zeiss Optical Works at Jena, Germany, in 1924. About 50 different lengths of iron rods were required, almost 4,000 rods in total, to create the 16 metre (52 feet) diameter dome, which was then finished in a thin layer of sprayed concrete. Thirty years later, in the United States, Richard Buckminster Fuller patented his own development of the principle and soon basic shells of timber-framed plywood panels (manufactured commercially) began to be erected, usually as dwellings.

Since 1954 more than 300,000 geodesic domes have been built for a variety of purposes, the most famous being Fuller's US Pavilion at Expo 67 in Montreal, which later burnt down.

The individual can take initiatives without anybody's permission.
RICHARD BUCKMINSTER FULLER (1972)

The EarthShip House
Taos, New Mexico, USA

MICHAEL REYNOLDS 1990

Architect Michael Reynolds has been developing self-sufficient housing made from recycled materials for more than 20 years. Apart from being environmentally-friendly, the houses also have the advantage of being very low-cost. In 1974 he built one for a judge in Taos utilising 75,000 beer cans. These were baled into 'bricks' of eight cans each, stacked up and then the walls were plastered to prevent them rusting. All for $15,000.

Garbage housing is a natural by-product of our throw-away society's conspicuous consumption and the EarthShip House combines the 1,300-year-old tradition of North American earthlodges with recycling contemporary detritus. Earthlodges were timber structures covered with sod or loose earth, then smeared with wet earth which dried like a plaster shell. The earliest were rectangular but later they became circular which added strength.

Here earth has been rammed into discarded steel-belted rubber tyres with a sledgehammer to produce a 136 kilogram (300 pound) adobe brick which is immensely strong. The walls are one metre (three feet) thick, a thermal mass which stabilises internal temperature at around 21°C (70°F). Aluminium beer cans provide infill and decoration.

The thermal mass of the house acts like a night-storage heater in reverse, absorbing heat during the hot days and re-radiating it during the cold nights. In more than 100 similar houses that he has built, Reynolds has also used bottles instead of cans, and panels of photovoltaic cells on the roof to convert sunlight into electricity for domestic power.

Tyres are as plentiful as trees on this planet, says the architect. We want to keep the latter but not the former, so let us use them as a building resource. 'If there weren't any discarded tyres we'd be inventing them to build with, because that's how good they are.'

'It just shows what can be done by taking a little trouble', said Eeyore. 'Do you see, Pooh? Do you see, Piglet? Brains first and then Hard Work. Look at it! *That's* the way to build a house', said Eeyore proudly.

A A MILNE *The House at Pooh Corner*

Favela Squatter Camps
Rio de Janeiro, Brazil

20TH CENTURY

I N Brazil they are called *favelas;* in Venezuela, *ranchos;* in Peru, *barriadas* or *pueblos jóvenes* – young towns; when built on stilts over water, in Brazil and San Salvador, *algados.* All the developing countries of the world have their equivalents, vast sprawling squatter camps on the edges of major cities, from Cairo to Karachi to Lima.

John F. C. Turner, an architect who spent 17 years helping squatter groups, estimated that perhaps half of all the housing in the world today is of this type; in Peru and several other countries, nearer two-thirds. In Brazil, First and Third Worlds meet in the short space between Copacabana Beach and the shanty towns overlooking the city. But the places offer more hope than it might seem at first. As Turner wrote: 'The ideal we should strive for is a model which conceives housing as an activity in which the users – as a matter of economic, social and psychological commonsense – are the principal actors'.

His experience was borne out by the 1976 study by Dr Janice Perlman, *Favelas: The Myth of Marginality,* which challenged conventional assumptions and showed that ordinary people were better equipped to resolve their housing needs than government agencies importing alien prefab high-rise systems from abroad, and at vast expense to the local economy.

'Houses in the beginning were very low, and like homely cottages or poor shepherd houses, made at all adventures of every rude piece of timber that first came to hand, with mud walls, and ridged roofs, thatched over with straw. But now the houses be curiously builded after a gorgeous and gallant sort, with three stories over one another.' Not Turner's or Perlman's words, but Sir Thomas More writing of his vision of Utopia more than 450 years ago.

Self-help housing provides the only viable and sustainable solution for tens of millions of people, as governments and city fathers have come to realise.

A CONTINUING TRADITION

This small almost finished house on its steep plot was my future, and whatever frustrations it held for me, every time I saw it I still experienced a small frisson of ... of what? Of love, I suppose, or something akin to that emotion.

WILLIAM BOYD *The Blue Afternoon* (1993).

Belvedere Village
Ascot, Berkshire, England

DEMETRI PORPHYRIOS 1989

THE Vernacular Classicism of Demetri Porphyrios shares deep roots with the work of 'contemporary regionalists' in terms of its response to climate, site and context, and in its use of simple forms. But while many of the latter employ the latest building materials and techniques – they describe their work as 'technologically progressive but culturally conservative' – Porphyrios prefers to use local and sustainable materials.

'As the century draws to a close it becomes clear that the debate on urban design and environmental policy is synonymous with architecture itself', he said. 'There must be a wholesale programme of reawakening and encouraging architecture that is beautiful, human in scale and which contributes positively to our cities and the places we create. In fact, the long-neglected wisdom of traditional building is showing fresh signs of life today, at a time when Modernism has proven incapable of delivering a socially and ecologically responsible programme of architecture or urbanism.'

Belvedere Village, situated on the Surrey/Berkshire border, comprises a small number of cottages, stables and farm buildings and is designed to have the simplicity and visual appeal of a typical English village. The layout is informal and a sequence of inter-related spaces pulls together the different buildings in a manner which is manifestly picturesque.

A principal element is the oak-roofed main barn with its dovecote tower, seen between two ranges of stables which have balconies to the grooms' accommodation above. The materials are stained softwood posts, brackets and rails, mixed red clay roof tiles and second-hand red Surrey and yellow London stock bricks.

Such is the patriot's boast, where'er we roam,
His first, best country ever is, at home.
OLIVER GOLDSMITH *The Traveller*

The Gypsy Vardo
England and Ireland

19TH CENTURY

THERE is no generic term for 'household' in Romany, the gypsy language, and in many parts of Europe gypsies still live in tents, caves, mud huts or small, single-room cottages. Others have adopted motorised homes. But in England and Ireland (where this picture was taken), the *vardo* – Romany for a living wagon – can still be seen occasionally, restored for holiday lets or confined to a museum exhibit. Less than 20 years ago it was estimated that there were still 600 families in Ireland living in *vardoes* – the gypsies' most treasured possession.

They used to be a common enough sight just after the First World War, when horse-dealing began to decline. Many were built by one company – Dunton & Sons, of Reading, Berkshire – whose name plate was a sure sign of quality.

There were six main types: Reading, Ledge, Burton, Pot-cart, Open-lot and Bow-top, the last being the most common for reasons of economy and the fact that its design made it the least likely to overturn. Typically it had a round canvas top on a bowed wood frame, with a chimney poking through the roof and no side windows. The internal layout was virtually identical: a bed space behind a partition at the far end of the central aisle; a corner cupboard, a seat and a chest of drawers to the right; and drawers, a stove with shelves and mirrors above, to the left. Interiors were often made in the finest french-polished mahogany. There was a door with removable steps at one end, usually at the front, and it was the custom for it to be drawn by one horse.

Sotheby's and other auction houses have now become involved in selling the finest caravans which once belonged to the kings of the road. In 1994 a Ledge covered in 22-carat gold leaf, and which cost its first owner £150 in 1915 – about as much as a semi-detached house at the time – had a reserve price of more than £40,000 at auction.

How does it feel
To be without a home
Like a complete unknown
Like a rolling stone?

BOB DYLAN *Like a Rolling Stone* (Song)

Hadrian's Villa
Tivoli, Italy

AD 118–34

THE Emperor Hadrian is best known in Britain for his military prowess and the 73-mile wall – Hadrian's Wall – which traverses the northern part of the country. At Tivoli, the ancient Tibur outside Rome, he is celebrated as an aesthete, philhellene and creator of one of the most ambitious and sophisticated of Roman palaces, Hadrian's Villa. The wall which bounded his estate was eight miles long.

In his *Letters* of around AD 100, Pliny the Younger had written persuasively of the delights of living in the country, where he owned two villas. Hadrian decided to emulate him but drew on his own extensive travels to plagiarise, in the most patrician sort of way, the architecture of other places within his empire and fill the result with looted Greek sculpture.

The villa was conceived on a vast scale, almost that of a small town, and he planted it down on an undulating site with magnificent views. The plan, made up of a series of axes which open up expansive vistas and create an element of surprise at each turn of a corner, responds to the location admirably in a conjugation of architecture, gardens, waterworks and statuary. Rectilinear and rounded masonry forms collide in a complexity of concave and convex shapes – a new type of spatial composition.

Here was Imperial Roman architecture in microcosm, with borrowed caryatids from the Erechtheion in Athens and an ornamental canal from ancient Egypt. Enough remains as ruins to get a good impression of Hadrian's eclectic taste as the magpie of classical antiquity.

Home wasn't built in a day.
GOODMAN ACE

The Alhambra, Granada Andalucia, Spain

1338–90

FORBIDDING red walls present a bleak face to first-time visitors to the Alhambra (from the Arabic *al-hamra'*, meaning red), an earthly paradise within a fortress high on a plateau. The palace had its origins in the eleventh century, but two North African rulers of Moorish Spain, Yusuf I and Mohammed V, made it their own, a princely pleasure dome whose roof is the sky. It is a mass of contrasts and contradictions: solids and voids; light and shade; taking on different guises by day and by night; a sequence of sometimes spacious, at other times intimate, spaces.

It was planned as a series of interlocking pavilions rather than conforming to an overall masterplan, in which shaded courtyards, water-channels, fountains and ponds, and lush gardens combined to seduce the senses. The decoration is elaborate: richly-carved honeycombed ceilings, horseshoe arches, large windows which are latticed or pierced, and vaults and capitals like serried ranks of stalactites (known as *muqarnas*), simultaneously dominant and recessive. The 'many-shafted fantasy' was too much for some to stomach, of course. For John Ruskin it was 'as beautiful in disposition as it is base in ornamentation'.

The Court of Lions, one of the two main courtyards set at right-angles to each other, measures an impressive 35 by 20 metres (115 by 66 feet) and is divided into four by water channels – an Islamic symbol of paradise. At its centre is an eleventh-century fountain bowl guarded by twelve white-marble lions, symbols of strength and courage.

Allusions to gardens are frequent in both the Bible and the Koran, although in the latter they tend to be more exotic – and erotic – in their appeal to the senses. By the beginning of the twelfth century almost 200 different species of flowers and plants had been catalogued by the Arabs and what is certain is that the colour and smell of the Alhambra's original garden, along with the sound of the trickling and gurgling water, would have fortified the Moors at play.

The Alhambra, the Acropolis, the Windsor Castle of Granada ... Few airy castles of illusion will stand the prosaic test of reality, and nowhere less than in Spain.

RICHARD FORD *A Handbook for Travellers in Spain* (1855, 3rd edition)

The Ponte Vecchio
Florence, Italy

Taddeo Gaddi 1345

The oldest of the five bridges that span the Arno in Florence, the Ponte Vecchio has held a special place in the affections of many writers, including Charles Dickens. In *Pictures from Italy*, published in 1846, the author best-known for his depictions of a rather squalid Victorian London was quite enchanted. 'The space of one house, in the centre, being left open, the view beyond is shown as in a frame; and that precious glimpse of sky, and water, and rich buildings, shining so quietly among the huddled roofs and gables on the bridge, is exquisite.'

From massive piers, the three segmental arches leap across the river in great bounds, almost oblivious to whether what they span is a summer trickle or winter flood of melting snow from the Apennines. Its name means 'old bridge' and such was its fame that during the Second World War it was the only one spared from destruction by the retreating German army. Gaddi, its designer, was a pupil of the painter Giotto for 24 years, although it is sometimes attributed to Neri di Fioravante.

The medieval use of bridges to accommodate houses and shops, as well as a thoroughfare, is unusual but not of course unique. Old London Bridge, begun in 1176 by Peter de Colechurch, soon had 100 houses along its length (at 282 metres – 926 feet – almost three times that of the Ponte Vecchio) as well as shops and a chapel. Its most famous was Nonsuch House, a four-storey building prefabricated in Holland then shipped over and held together with wooden pegs. Three times all the buildings were destroyed by fire and rebuilt, until they were finally demolished in 1754.

On the Ponte Vecchio, the original butchers' shops were replaced by jewellers' in the sixteenth century, and it was dubbed the Bridge of the Goldsmiths. Many Renaissance artists and architects trained as goldsmiths, including Brunelleschi, but the houses on the bridge today are the result of centuries of gradual alteration, accretion and demolition.

Taddeo Gaddi built me. I am old
    Five centuries old. I plant my foot of stone
Upon the Arno, as St Michael's own
    Was planted on the dragon. Fold by fold
Beneath me as it struggles, I behold
    Its glistening scales.

HENRY WADSWORTH LONGFELLOW
The Old Bridge at Florence (1874)

Little Moreton Hall
Cheshire, England

LATE 15TH TO LATE 16TH CENTURY

L OOKING like 'some vast unstable doll's house', Little Moreton Hall is the most picturesque half-timbered house in Britain. Modified from an original late fifteenth-century hall, its construction spans the period from early-Tudor to early-Elizabethan but is very much in the medieval tradition.

Its walls and gables seemingly defy gravity; its roof sags under the weight of its gritstone slates; and yet the fact that it has survived intact for more than four centuries is a testament to the sound techniques as well as to the grandiloquent aspirations of the Moreton family and their carpenter Richard Dale. A riot of craftsmanship can be seen in its oriel and bay windows, its beams in diamond and quatrefoil patterns, its moulded mullions and transoms, while internally the ceilings boast curved braces and emblematic panels of plasterwork. Massive brick chimney breasts and buttressing, and the simple stone bridge which provides the only access across the moat, come as something of a light relief by way of contrast.

When standing in the cobbled courtyard and looking up, the most impressive feature of all is the long gallery which was added – almost as an afterthought? – right across the top at roof level. It measures 23 metres long (75 feet) by less than 4 metres wide (12 feet 6 inches) and features a continuous band of windows both north and south along it. With mullions separating the narrow areas of glass, these windows, like those below, form an integral part of the timber-frame structure.

Plaster decoration within the gallery depicts the Spear of Destiny whose rule is Knowledge and the Wheel of Fortune whose rule is Ignorance. With hindsight, one might observe that the Moreton family put their faith in the former rather than succumbing to the latter.

Little Moreton Hall is the most popular of all English black and white houses, yet as it comes into sight – happily reeling, disorderly, but no offence meant – it seems at first unbelievable, and then a huge joke.

NIKOLAUS PEVSNER and EDWARD HUBBARD
The Buildings of England: Cheshire (1971)

Villa Rotonda
near Vicenza, Italy

ANDREA PALLADIO 1567–9

GOETHE called it 'habitable but not homely'. Lord Burlington, more than a century-and-a-half after the original, built his own version of it – Chiswick House in London. Villa Rotonda (also known as Villa Americo-Capra, or simply Villa Capra) was one of the most influential examples of domestic architecture to come out of Renaissance Italy, and with good reason.

For the first time equal status was given to all four elevations, a decision taken by Palladio on the basis that the hilltop views in every direction deserved equal attention. What's more, he chose projecting temple porticos with which to do it on all sides of the centralised square plan. A dome, usually associated with churches and cathedrals, was used to denote the round form of the salon at its centre, a device which had been used in Renaissance houses before but not so that it projected above the roofline. In a word, it was unique.

Although the villa served as a rural retreat, so imbued was the architect with recreating the classical grandeur which went hand-in-hand with symmetrical planning and the use of harmonic proportions (which relate architecture to music and the harmony of the universe) that he explained his concept in these terms: 'The handsomeness will arise from a fair form, and the correspondence between the Whole and its Parts, of the Parts among themselves, and of them to the Whole: because of that a Building ought to appear an entire and perfect Body, wherein each Member agrees with the others, and all the Members be necessary to what you design'.

Born Andrea dalla Gondola, Palladio was the nickname given to the architect by his tutor after Pallas, a title of Athena – goddess of wisdom – in classical mythology. In turn his new name gave rise to a style of architecture derived from his own buildings and books, Palladianism.

In fact there was but one thing wrong with the Babbitt house; it was not a home.

SINCLAIR LEWIS *Babbitt*

The Queen's House
Greenwich, England

Inigo Jones 1616–35

INIGO Jones is sometimes described as England's first architect. He wanted an end to the practice whereby builders worked to pattern books or a few sketches, and for architects to assume total responsibility for the design and execution of their work. He was also familiar with the buildings and writings of Andrea Palladio in Italy, and was determined to introduce Classicism to England. In 1615, when aged 42, he was given the influential appointment of Surveyor-General to the king and the following year began to design what was to be his most influential commission.

James I gave Greenwich Palace and Park to his wife, Queen Anne of Denmark, and Jones was asked to design for her a new residence – more a hunting lodge than a home. The Queen died before it was completed and the estate was inherited by his son, later Charles I. He in turn ordered it to be completed for his wife, Henrietta Maria.

Palladio would have approved of Jones' design: symmetrical about each axis, the two blocks either side of a road were joined by a bridge at the centre at first-floor level. Later additions by John Webb turned the H-shaped plan into a complete rectangle.

A galleried hall – a perfect cube just over 12 metres (40 feet) in each dimension – faces the river and then gives access to a loggia overlooking the park. The elevations are divided into three, in the Palladian manner, and have large sash windows, porticoes, pediments and balustrades. Today the house is the centrepiece of the National Maritime Museum.

The Royal Naval College (formerly the Royal Hospital) by Jones' direct architectural heirs, Wren, Hawksmoor and Vanbrugh, sits either side of the central axis leading down to the Thames. In the distance, just to the right of the axis, is the Royal Observatory, built in 1675 by Wren. The overall view has rightly been described as 'the finest architectural panorama in England – some would say in Europe'.

Discriminating observation of the past must be the inspiration for the future.

HRH THE PRINCE OF WALES *A Vision of Britain* (1989)

Palace of Versailles
Yvelines, France

LOUIS LE VAU AND JULES HARDOUIN-MANSART 1669–82

THE *roi soleil* did not much care for Paris, as the citizens of Paris did not much care for him. Most of his time was spent touring his various châteaux around the capital until he decided, in 1661, to move his whole court and government to Marble Court, a small hunting château built by his father at Versailles. Within 20 years it had become a vast dormitory suburb accommodating 10,000 people, and Louis XIV only visited Paris again four times in the last 22 years of his life.

He employed Le Vau, the leading Baroque architect of the day, who upon his death in 1670 was succeeded by Hardouin-Mansart. Le Brun was the decorator; Le Nôtre the garden planner. More than 30,000 workmen were involved in constructing the amazing pile, whose principal façade is 580 metres (1,902 feet) long and has 375 windows. Later Louis XV added the first passenger lift to give private access to his mistress on the floor below. The king's playground was set in 9,700 hectares (24,000 acres) of gardens with a mile-long Grand Canal and fountains fed by water pumped from 10 miles away.

By 1664 the gardens were sufficiently advanced to stage a fête, the like of which has never been seen before or since. It went on for days. Musicians dressed as shepherds wandered through the crowds. Hedges and trees were clipped to represent the different orders of architecture. Water in the fountains was scented. At night the whole vast estate was lit up and Chinese fireworks were set off, while the man around whom all this revolved received his guests dressed in silver armour studded with diamonds.

The extravagance and remoteness of the king and his successors helped to precipitate the French Revolution a century after the palace was nearing completion, and the garden was only saved by turning it into a vast allotment to grow food for the people. Surveying Louis XIV's Versailles, Madame de Maintenon remarked: 'Apart from passion, I have never seen anything more sad'.

'Sire – over what do you rule?'
'Over everything', said the king, with magnificent simplicity...
For his rule was not only absolute: it was also universal.
'And the stars obey you?'
'Certainly they do', the king said. 'They obey instantly. I do not permit insubordination.'

ANTOINE DE SAINT-EXUPÉRY *The Little Prince* (1945)

The Winter Palace
St Petersburg, Russia

V. V. Rastrelli 1754–62

As the home of the tsars from Peter the Great, in 1704, to Nicholas II, in 1917, the Winter Palace has seen more than its fair share of history. Today, where once Rasputin wielded his hypnotic power over the Imperial family, its glittering suites crammed full of paintings and other works of art form part of the Hermitage Museum and throng with Russian and foreign visitors.

The present palace, the fourth, was built in the Russian Baroque style by Rastrelli 'solely for the glory of all Russia'. It contains more than 1,000 rooms around an inner courtyard and within sea-green and white walls of powerful plasticity fronting Palace Square. Completed in 1762 under Catherine the Great, it became the setting for the finest collection of Western European art from the thirteenth century onwards. But as the journalist Alexander Herzen wrote at the time, it was 'like a ship floating on the surface of the ocean, it had no real connection with the inhabitants of the deep, beyond that of eating them'.

Peter the Great was never happy living in the previous palace's large rooms, preferring a small retreat 25 miles outside the city; Nicholas I took a barrack-like apartment within it; Alexander II made his home in one small corner; and Nicholas II took refuge from the increasing unrest outside in a series of attic rooms.

It played a key role in the Revolution of 1905–7, and again in the October Revolution of 1917 when looting was happily confined to the wine cellars. During the Second World War some two million works of art were evacuated to the Urals for safekeeping. But now, wrote Alexander Creswell, who painted the watercolour from which this detail is taken, 'beneath the shuffling feet, uncounted works of art stand on duckboards as the river Neva seeps into uncharted cellars'.

Small rooms or dwellings discipline the mind, large ones weaken it.

LEONARDO DA VINCI *Notebooks* (c.1500)

No.13 Lincoln's Inn Fields London, England

Sir John Soane 1812–3

Born and brought up in Goring-on-Thames, Oxfordshire, Soane was apprenticed to George Dance the Younger, Surveyor to the City of London, when he was 15, and later travelled to Italy on a scholarship from King George III. Those early experiences, combined with his admiration for the work of Hawksmoor and Vanbrugh, prepared him well for an illustrious career as an architect and academic. Fate dealt him a cruel hand, however, and unlike the prodigious output of his contemporary and rival, John Nash, few of his London buildings remain.

His highly personal interpretation of Neo-Classicism – using abstract geometry and simplified lines – was first demonstrated in country houses before being given full reign when he was appointed architect to the Bank of England at the age of 35. Only an outer screen wall now remains of that extensive rebuilding he undertook; and his Dulwich College Picture Library received a direct hit from a flying bomb during the Second World War, although it was subsequently rebuilt.

But his ingenuity in devising comparatively large and complex interiors found ample expression on a domestic scale in his home at Lincoln's Inn Fields, now the Soane Museum. There are in fact three separate dwellings: No.12, built in 1792 for himself and his family and with his office at the back; No.13, built 20 years later as a domestic museum for his varied collections; and No.14, built in 1824, which was let to tenants but has one of the museum galleries running along the back.

He used shallow domes and curving vaults with top lighting to dramatic effect in a series of interconnecting rooms, some large, some small. Reflections from strategically placed mirrors and ceiling bosses, and the manner in which vistas were opened up and then terminated, make this a home full of, in his own words, 'hazard and surprise'.

Light, God's eldest daughter, is a principal beauty in a building.

THOMAS FULLER (1642)

Cumberland Terrace, Regent's Park, London, England

John Nash 1827

As the British empire was beginning to reach the height of its power in the first half of the nineteenth century, the architect John Nash was busy creating what no-one else had the energy – or nerve – to take on: a route through central London which would link Buckingham Palace, Carlton House and the Mall eastwards towards the Strand, and northwards through Piccadilly and Regent Street to Portland Place and, eventually, Regent's Park. Much of his plan and many of his buildings still survive today, although much has also been lost in subsequent waves of property speculation.

For as Mark Girouard recalled, Nash, as well as being a very talented 'if rather flashy' architect, was also a speculator in his own right. He would often produce only a rough sketch of what he envisaged and left his assistants to detail and gerry-builders to execute. But stucco can hide a multitude of shoddy brickwork to theatrical effect.

Cumberland Terrace is one of his most exuberant legacies. House lots in London were narrow, often little more than 6 metres (20 feet) wide if more like 30 metres (100 feet) deep. So following the French manner, Nash dressed up his terraces as if they were a single mansion or palace, and the newly-emerging merchant and middle classes appreciated such ostentation.

The three blocks, linked by arches and with courtyards behind, contain 31 houses. There is a ten-column portico with pediment and allegorical sculptures overlooking the park and the site of what was intended to be a new house for the king, which was never built. In this case his client was the Commissioners of Woods, Forests and Land Revenues to whom he was the official architect.

Sighed JOHN NASH to DECIMUS BURTON
'Fashions change, but of one thing I'm certain:
    Our classical terraces
    will be treated as heresies
e'er the stucco's had time to get dirt on.'

P E C Anti-Uglies: Some highlights of architecture (1959)

Maison Horta
Brussels, Belgium

VICTOR HORTA 1898–1901

THE flowering of Art Nouveau was short-lived, but its impact was all the greater for distilling so much creativity into so little time. At the Turin exhibition of 1902, the Italian critic Silvius Paoletti rejoiced in the way it had taken the place of 'pitiless authoritarianism, rigid and regal magnificence, burdensome and undecorated display'. Instead, the movement offered 'delicate and intimate refinement, fresh freedom of thought, the subtle enthusiasm for new and continued sensations … art has new aspirations, new voices and shines with a very new light'.

Its most successful exponent in Belgium was Victor Horta. Virtually all his important buildings were executed in a space of 10 years, mostly narrow-fronted houses on the new suburban boulevards of Brussels, which were typically just three-and-a-half metres (11 feet) to six metres (20 feet) wide.

Street elevations commonly had gently curved lintels over the windows and a profusion of wrought-iron balconies with the familiar whiplash curve. But the exteriors of his houses were modest by comparison with the interiors which were highly expressive statements of structure and the use of contrasting materials.

In his Hôtel van Eetvelde for a wealthy baron, slender iron columns ring a superb octagonal space at the centre of the house under a shallow glass dome, which allows light to flood the interior. In his own house, and on a more modest scale, wrought iron, wood, and clear and stained glass are combined in florid forms taken from nature, so that each room with its specific function flows seamlessly into the next, and is complemented by sinuous surface ornament.

All architecture is shelter; all great architecture is the design of space that contains, cuddles, exalts or stimulates the persons in that space.
PHILIP JOHNSON

Artist's Cottage and Studio Farr, Inverness, Scotland

CHARLES RENNIE MACKINTOSH 1900; BUILT 1990–2

CHARLES Rennie Mackintosh only achieved cult status long after his death. Although his Glasgow School of Art, the result of an architectural competition in 1896, is now recognised as among the finest buildings of its period anywhere in the world, he was better appreciated in Vienna than in London or his native city. He also built few houses. The Artist's Cottage is thought to have been designed by the architect as an idyllic rural retreat for himself and his new bride, Margaret Macdonald, but without a specific site in mind.

Ninety years later Dr Peter Tovell, a Mackintosh enthusiast, thought it would make an ideal home for his wife Maxine and their two children, on a site of their choosing. They commissioned Robert Hamilton MacIntyre to execute the design as if he were the job architect. New drawings were prepared by scaling off the originals in the Hunterian Art Gallery at the University of Glasgow, and by consulting Professor Andy MacMillan.

Externally the cottage is built as envisaged, with its battered walls in traditional Scottish roughstone harling, decorative wrought-iron surrounds and balcony, slate roof and leaded glass windows, and solid vertical chimney stacks. Internally, however, 'the ratio of hearths to toilets has been inversed': ideas about comfort and the affordability of domestic service have changed dramatically in the intervening period. Most of the ten chimneys are not used and underfloor electric heating has replaced the inconvenience of real fires; and instead of the two lavatories intended, there are now five en-suite shower rooms, a separate bathroom and a cloakroom.

Local artists were employed to replicate Mackintosh's leaded glass panels and ironwork, and replica Mackintosh furniture is to be found in the dining-room and elsewhere. One original Mackintosh armchair recently sold for £300,000 – more than the current market value of the house.

All architecture is great architecture after sunset.

G K CHESTERTON

The Schröder House Utrecht, Holland

Gerrit Rietveld 1923–4

The neighbours were understandably shocked when the house Mrs Schröder-Schräder commissioned and helped to design with Gerrit Rietveld suddenly appeared at the end of their conventional brick terrace, 70 years ago. While the architect succeeded in his aim of using 'elementary forms, simple spaces and primary colours exclusively, because they are so fundamental and because they are free of associations', it must have been an unwelcome and alien intrusion.

Rietveld was originally apprenticed in his father's joinery shop and he spent many hours refining models of the house before he was happy with the result: a pure, abstract geometry defined by planes which slot together as if made with a pack of playing cards. In its plans and elevations it could be a three-dimensional painting by Mondrian – the leading member of the De Stijl movement, about whom Louis Hellman wrote the memorable clerihew 'Mondrian/What a man/Framed a chart/Called it art'! Rietveld was also a prominent member of the group.

Internally, the ground floor is sub-divided so that each room can be used independently of all the others. Upstairs, in what was known as the Loft, one large space is subdivided by means of sliding and demountable partitions. As the walls project and recede alternately, blurring inside and outside, one is reminded of the influence on De Stijl of Frank Lloyd Wright's houses and, indeed, the 1,000-year-old tradition of Japanese screen houses.

As an inhabited piece of sculpture the house has few equals. But looks can be deceptive: much of what appears to be poured concrete is in fact plastered brick and timber. Perhaps it should be considered as a sham, a modern folly? But that would be to deny its extraordinary influence on a generation of architects. Recently restored, it stands as one of the most potent icons of the Modern Movement.

The evolution of culture marches with the elimination of ornament from useful objects.
ADOLF LOOS

Maison de Verre
Paris, France

PIERRE CHAREAU AND BERNARD BIJVOET 1928–32

THE socialist writings of the poet Scheerbart could have been the starting point for Pierre Chareau when he was commissioned by Dr and Madame Dalsace, a gynaecologist and his wife, to design a house and surgery for them underneath an existing eighteenth-century house. 'Glass brings us a new age', Scheerbart wrote. 'Brick culture only does us harm.'

Whatever constructional drawings may have been prepared for this innovative house they have never been found; it is more likely that rough sketches sufficed as the architects worked for four years with the client and a steel fabricator, Dablet, to underpin the house above it and fill the walls with translucent bricks of glass.

Inside there is an ingenious layout of reception and consulting rooms, on the ground floor, with a double height salon, study, kitchen, dining- and sitting-rooms, bedrooms and bathroom above, connected by three separate staircases. Chareau, who was primarily a designer and furniture maker, fitted it out with sophisticated sliding and pivoting doors, partitions and cupboards.

The arrangement of rooms ensures privacy for the surgery and the massive glass walls allow light to flood in while neighbourly intrusions cannot. At night, external floodlights create a warm glow inside the living space.

Le Corbusier was a frequent visitor to the house while it was under construction and no doubt he marvelled at this early example of a 'machine for living in'. Many others have been influenced by it since, from Charles and Ray Eames to Michael Hopkins, Sir Norman Foster and, especially, Sir Richard Rogers.

And before the throne there was a sea of glass like unto crystal.

REVELATIONS chapter 4, verse 6

Pacific Palisades
Santa Monica, California, USA

CHARLES AND RAY EAMES 1945–50

IT MAY look prosaic, but as Reyner Banham said, it was 'one of the touchstones of recent architecture'. Pacific Palisades, better known among the *cognoscenti* as simply 'The Eames House', began its life as Case Study House No. 8 – one of a series of houses commissioned, designed and built at the behest of *Arts and Architecture* magazine between 1945 and 1962. The houses were published in the popular journal and its readers were invited to take a look at the finished work.

The Eames distinguished themselves as, first and foremost, designers of avant-garde chairs, in moulded plywood, wire and plastic, and cushioned leather. For many, Charles was a latter-day Renaissance Man, a film-maker of exceptional talent and designer of toys and exhibitions as well. So when it came to producing an isolated villa in the suburbs of Los Angeles, interest in the result was not confined to close neighbours.

Technology was given the upperhand in creating what is, in effect, a one-off, off-the-peg house – bespoke insofar as it was tailor-made to their own needs, but off-the-peg in terms of its conception and assembly. For it served not just as a place in which to live but as a display case for their furniture collection and assortment of *objets d'art* and *objets trouvés*.

The double-height living spaces, and studio in a separate block across a small courtyard, are enclosed by a steel frame with exposed open-web steel trusses spanning the six metre (20 feet) width. Corrugated steel decking, also exposed, is the ceiling and roof. It was all erected in a week. The glazed and solid panel walls are a Mondrianesque pattern of standard factory windows ordered from a catalogue. It has the feel of a Japanese timber frame-cum-lightweight screen house, and its 'industrialised vernacular' found followers in Britain's post-war school-builders and – like the Maison de Verre – among a generation of architects, including Sir Norman Foster and Michael Hopkins.

Architecture, unlike other arts, is not an escape from, but an acceptance of, the human condition, including its many frailties as well as the technical advances of its scientists and engineers.

PIETRO BELLUSCHI

The Glass House, New Canaan Connecticut, USA

PHILIP JOHNSON 1949

BUILT for himself, Philip Johnson's Glass House is probably the most famous American house after Frank Lloyd Wright's Fallingwater, 'an ageless humanist dream', according to the critic Vincent Scully. The architect's stated sources are many and sometimes contradictory but include Mies van der Rohe, Schinkel, Choisy, Ledoux and the artist Ben Nicholson. The design has rightly been celebrated for the precision in its use of materials, its clarity and its siting within the architect's own private estate. By day it is a glass jewel box; by night, the walls disappear and it resembles an open-sided Japanese pavilion as the floodlit surroundings reclaim the ground on which it rests.

The house is essentially one large room all on one level, a plinth of herringbone patterned brick, from which black-painted steel columns rise to the roof. The walls are plate glass and only the bathroom is enclosed, in a solid drum which rises through the roof, and which has a fireplace set into its convex outer wall facing the living space. A low kitchen unit is placed immediately to one side of the entrance, the rest is left entirely open apart from a minimum amount of furniture, a potted plant and a sculpture. The atmosphere is pristine, elegant and tranquil.

This bachelor's weekend retreat may be likened to a latter-day Thoreau's Walden Pond out in the woods, or even a domesticated Greek temple. On a more mundane level, the architect said that it was like living amidst nature, 'without having to get into one of those horrible sleeping bags'. But in the midst of winter, with logs burning on the fire, the lights out and the snow falling, the poetry of the place returns: 'it is almost like a celestial elevator'.

'And when I found the door was locked
    I pulled and pushed and kicked and knocked.
And when I found the door was shut,
    I tried to turn the handle, but ...'
There was a long pause
    'Is that all?' Alice timidly asked.
'That's all', said Humpty Dumpty. 'Goodbye.'

LEWIS CARROLL *Through the Looking-Glass*

Byker Wall
Newcastle-upon-Tyne, England

RALPH ERSKINE 1969–81

COMMUNITY architecture, the process by which the active participation of end-users leads to a more responsive end-product in terms of their needs and aspirations, is timeless, although the term was only coined in its present sense (by the author) in 1975. Six years earlier, three unrelated events conspired to make it one of the most significant movements of the late twentieth century.

The first was the publication of the Skeffington report, *People and Planning*, commissioned by the British government, which advocated public participation in planning. The second was Shelter's Neighbourhood Action Project (SNAP) in Liverpool, in which architects lived in the area where they worked and were answerable to local decisions. The third was when Ralph Erskine was appointed to undertake the redevelopment of Byker and also set up a site office in the midst of his 'real' clients, in a disused funeral parlour, and operated an 'open-door' policy.

Byker has three distinctions among major social housing projects in Britain: it is popular with virtually all those who live there; active participation by its residents and pride in their surroundings has reduced crime and vandalism, and led to a Best Kept Village Award; and Erskine received the approbation of his professional peers by being given their ultimate accolade, the Royal Gold Medal, in 1987.

More than 2,300 new dwellings were created on a slum-clearance area, 80 per cent of them in low-rise terrace housing, the balance in a huge 'wall' of three to eight storeys more than half a mile long, which screens the rest from noise pollution and winds from the North Sea. The brief was continually modified to reflect local opinions. The estate's scale and detailing are informal, and residents continue to lavish care on it.

The result, said critic Charles Jencks, is 'the most humane public housing in the world'.

The community architecture movement represents an important element in the drive by people for more say in the decisions that shape their lives. It offers the opportunity for democratic action and should help to renew the strength of our democratic processes.

LORD SCARMAN
Foreword to *Community Architecture* (1987)

Douglas House, Harbor Springs, Michigan, USA

RICHARD MEIER 1971–3

RICHARD Meier came to prominence as one of the so-called 'New York Five', also known as 'The Whites' due to the whiteness of their buildings. They took a rigorously intellectual approach to their art and much of their energy was devoted to teaching at East Coast universities. They also considered the early years of the European pioneers of modern architecture as 'some lost Golden Age', in the words of critic William Curtis.

But a deftness of touch replaced the rather more ponderous formalistic statements in concrete which inspired them, no more so than in Meier's concern for vertical rather than horizontal planes, circulation routes being used to generate the plan, transparency, and the use of thin structural elements, often in metal. His houses combine a simplicity of concept with a complexity of execution and achieve considerable elegance as a result.

The 'whiteness' of the solid structure and reflections of the glazed voids of the Douglas House make a stark contrast to the steeply wooded hillside overlooking Lake Michigan, its isolation reinforced by the subsequent designation of the neighbouring shoreline as national seashore which prohibited further houses nearby.

Approached at roof level by a flying bridge, the five-storey house has two means of vertical circulation, one internal and one external. A curved skylight allows natural daylight to penetrate deep into the heart of the dwelling, a device borrowed from Italian and southern German Baroque architecture. On the entrance side, walls are solid, denoting 'private' spaces; on the lakeside, walls are glass, giving magnificent views out and denoting 'public' areas, including a double-height living-room.

The cantilevered external staircase and stainless steel chimneys add to the overall impression that this is a modern Noah's Ark beached high up on Mount Ararat after the flood had receded.

In architecture the pride of man, his triumph over gravitation, his will to power, assume a visible form. Architecture is a sort of oratory of power by means of forms.

FRIEDRICH WILHELM NIETZSCHE *Skirmishes in a War with the Age, Twilight of the Idols* (1888)

The Hopkins House
Hampstead, London, England

Michael and Patty Hopkins 1975–6

THE Hopkins House clearly owes a debt to the steel-and-glass pavilions of Mies van der Rohe and his acolyte Philip Johnson, as well as to Pacific Palisades by Charles and Ray Eames. But this house does not sit in splendid isolation but amidst the Georgian and Regency villas of a Hampstead street.

Michael Hopkins, who worked with Sir Norman Foster, established his own practice in the mid-1970s and needed an office as well as a family home. The extent of the building was easily defined by building lines and the topography of the site, which is three metres (10 feet) below road level. The entrance to the simple rectangle, which measures 10 metres (33 feet) by 12 metres (39 feet), is at first-floor level via a footbridge, with what was originally an office on the upper level of the two storeys, and an open spiral staircase near the centre of the house providing vertical circulation.

The house was in fact used as a test-bed for developing ideas for larger commercial buildings. Only the bathroom and store are enclosed by fixed partitions of prefabricated melamine-faced panels; the rest is a slender frame of steel with plate glass infill panels (50 per cent of which can be opened as sliding windows), with open and exposed lattice trusses and open corrugated metal decking for the ceiling and roof. The flank walls are of insulated heavy steel decking.

Within the open-plan spaces of both levels, different functions are defined by aluminium venetian blinds, which also provide some degree of privacy from passers-by in the street. In strong sunlight the blinds produce a striped chiaroscuro across the carpet, furniture and other surfaces. The effect is light and airy and the house has become one of the major reference points in the High-Tech tradition of modern British architecture.

The house of cement, iron and glass, without carved or painted ornament, rich only in the inherent beauty of its lines and modelling, extraordinarily brutish in its mechanical simplicity, as big as need dictates, and not merely as zoning rules permit, must rise from the brink of the tumultuous abyss.

SANT'ELIA *Messagio* (1914)

Self-Build Housing
Lewisham, London, England

Walter Segal 1977–87

Every year since 1984 in Britain, more than 10,000 houses have been self-built. This phenomenon is not quite as remarkable as it might at first appear, given that man has an instinct to build (even if it's only a weekend's devotion to DIY), that the market cannot satisfy everyone's personal preferences and idiosyncrasies, and that the savings to be made can amount to between 20 and 40 per cent of the cost of a conventional house.

Access to a suitable plot, the availability of financial resources and the advice of an architect or surveyor, are the principal ingredients, along with a considerable commitment of time and energy on the part of the builder, what the Americans call 'sweat equity'. Many would argue that self-build is equally appropriate in a post-industrial world as in the pre-industrial era.

Walter Segal devised a sophisticated timber-frame system, which became known as the Segal Method, enabling people with no previous building experience – including old age pensioners assisted by their grandchildren – to create their own homes with hand-held tools and a simple manual. The method appealed not only to the self-builders but to a younger generation of architects who felt detached from the actual process of building. The evident pride of those who have taken part recalls the words of Abraham Lincoln: 'I like to see a man proud of the place in which he lives; and so live that the place will be proud of him'.

Self-build houses provide choice, flexibility and adaptability. Easily-worked, standardised materials are used with maximum economy, 'wet' trades such as bricklaying and plastering are eliminated, and specialised skills are only required for roofing and services. The role of the adviser changes, significantly, from being an expert 'provider' to that of being an expert 'enabler', helping people to solve their own problems rather than coming up with all the solutions.

A home is not a mere transient shelter; its essence lies in its permanence, in its capacity for accretion and solidification, in its quality of representing, in all its details, the personalities of the people who live in it.

H L MENCKEN *Prejudices*: Fifth Series (1926)

Eagle Rock
Uckfield, Sussex, England

Ian Ritchie 1982–3

EAGLE Rock is a lightweight bird that will never fly, despite its aerodynamic, symmetrical profile perched on a secluded site – discovered through the pages of *Exchange and Mart* – from which it takes its name. An existing bungalow was unceremoniously blown up and 15 tons of rock were removed before construction could begin.

It was originally conceived as a piece of High-Tech sculpture, its structure painted in black with red for the steel joints, but it will soon take on the natural camouflage of its surroundings and virtually disappear from view. Trellises within which it is wrapped, and the inclined steel columns, struts and wires which hold up its 'head' and 'beak', will be the greening of its architecture.

Ian Ritchie's client was a collector of rare orchids. At one end of the house, the 'beak', is an extended entrance canopy; at the other end, the 'tail', a conservatory. In the main body of the house is the living space. Each 'wing' has a bedroom suite, on one side for the owner, on the other for guests. The walls are either sliding glass or sand-coloured cladding (covered by the trellises) and the floor of the house is the same colour as the surrounding paving.

High-Tech design and low energy consumption are rarely combined, but here the collection of passive solar energy was a prime consideration. The house is double-glazed and insulated to a high specification; the louvre blinds are also insulated and solar collector panels are arrayed on the roof.

Architecture is not the answer to the pragmatic needs of man but the answer to his passions and imagination.
EMILIO AMBASZ

The New House
Wadhurst, Sussex, England

JOHN OUTRAM 1983–6

JOHN Outram's idiosyncratic approach to architecture is nowhere better displayed than in this modern English country house, set in a landscaped deer park, which fuses contemporary technology and materials with historicist references to create a work of stunning exuberance. The fact that it is impossible to classify in stylistic terms is no bad thing as that is often the measure of true originality.

Different camps claim it for their own: Classicists for its reinterpretation of the Palladian villa; Modernists for its structural steel frame and concrete cladding; Post-Modernists for its scavenging of previous architectures. If there were contemporary Gothicists (Outram designed the recent Pugin exhibition at the Victoria and Albert Museum) they would no doubt point to his use of polychromy and adherence to Pugin's dictum that 'all ornament should consist of enrichment of the essential construction of the building'.

Stated influences include ancient Egyptian hieroglyphics, Minoan Greece, Alberti, Le Corbusier and Louis Kahn, and the graphics of Cassandre. But much of the decoration, especially of the Classical details, is self-confessedly 'ad hoc'. His aim is to 'reconnect architecture to its natural roots'.

No surface or detail is considered unworthy of embellishment. The floor of the entrance hall is a marbled pattern based on a section through fluted columns, blown up to Brobdingnagian proportions. Shallow ceiling vaults are reminiscent of Sir John Soane, but made in bent plywood. The semi-cylindrical brick columns which frame the new entrance to a Victorian greenhouse support an oversized entablature. Doors and windows are inlaid with trellice-work. Even the concrete cladding is made up of four different and identifiable types of aggregate, including 'Blitzcrete' – a mixture of crushed brick and rubble which is ground and polished to reveal chips of terracotta as if it were some precious stone.

So, from day to day, and strength to strength, you shall build up indeed, by Art, by Thought, and by Just Will, an Ecclesia of England, of which it shall not be said, 'See what manner of stones are here', but, 'See what manner of men'.

JOHN RUSKIN *Lectures on Art* (1870)

Alpine Houses
Switzerland

*c.*2,000 BC

CHARLES Dickens called Switzerland 'the land of wooden houses'. Sixty years later Ernest Hemingway, in a sample of his inimitable journalism, recalled it as 'a small, steep country, much more up and down than sideways, and all stuck over with large brown hotels built in the cuckoo-clock style of architecture'. What both refer to is a style which developed from the Alpine cabins of the Middle Bronze Ages almost 4,000 years ago, and which was intended to keep the harsh climate at bay.

Then logs were stacked up on stone bases under triangular roofs of wooden slates, called shingles, which were weighted down with stones – a practice still common on the low-pitched roofs with wide eaves of the Bernese Oberland. Away from the high valleys, roofs are usually steeper and often have dormer windows. Projecting weatherboards at each storey shelter ribbon windows. Access to the ground floor, almost a storey itself above street level, is often by means of an external staircase to the leeward side. Timber window shutters are hinged vertically and open sideways; or are at the top and open upwards. Fretwork is a matter of pride and joy, on beams, brackets holding up the overhanging eaves and exposed rafters and bargeboards.

In the Valais, the French-speaking region, the houses are often much taller with open, projecting timber galleries running alongside the timber living spaces, connecting them to a stone-built kitchen block.

Apart from the Swiss cottage or chalet (little château), Switzerland's most famous architectural export was Le Corbusier whose style was anything but cuckoo-clock.

The strength of a nation is derived from the integrity of its homes.

CONFUCIUS

Painted Timber Shacks
Eastern Caribbean

Taxophilia, the love of classification, has been extended in recent years to cover the vernacular architecture of virtually every region of the world. As tourism, the world's largest industry, has spread its tentacles to ever more remote parts, instant satellite communication has made the 'global village' a reality, and Coca-Colonisation has created global markets, uniqueness has come to be prized as an asset. Even the relatively humble, simple and unambiguous shacks of the Eastern Caribbean have been analysed, catalogued and subjected to measured drawings, which has thrown up at least a dozen variations on the theme.

The typical house of the region, known as a *case,* is based on a single room even if it houses an entire family, because life is lived for the most part out-of-doors. Larger houses have two or three rooms, with a kitchen and bathroom in a separate but adjoining structure.

Timber is the most common building material, with corrugated iron, timber shingles, tiles or even bricks for the pitched roof. Some houses are transportable, almost all are brightly painted; and many are perched on concrete piles with an open staircase leading to the front door, or a solid stone foundation, to prevent the transfer of humidity from the soil.

Protection from direct sunlight is afforded by overhanging eaves or hoods to the windows, porches and galleries, and louvred shutters to the windows. Attics are also ventilated for storage. Symmetrical façades, with windows either side of the front door, are a common feature of islands colonised by the British, as shown here in Antigua.

[The vernacular] does not go through fashion cycles. It is real, immutable, indeed unimprovable, since it serves its purpose to perfection.
BERNARD RUDOFSKY

The Doges' Palace Venice, Italy

1340 AND 1424 (Rebuilding)

THERE is a Hindustani proverb which states that 'an arch never sleeps'. No more so than in a Gothic arch, and in the way it is employed in the Doges' Palace, one of the finest examples of its type in the world. For seven centuries it has been at the heart of the city, first as a square fortress with defensive walls and corner towers, but later transformed into the airy palace we see today after a series of fires and rebuilding programmes.

As well as providing the official residence of the Doge, elected for life like the Pope, the palace also served as an armoury, courtrooms and dungeons, and within it was built the chamber of the city's ruling Grand Council. But it now serves as a living museum of Venice's history and its artists.

The light arcade at Piazzetta level, which used to be two steps higher, supports an even lighter tracery of the gallery above and then a sheer mass of patterned wall, around bay windows and further arches of the windows. The whole is topped with spiky pinnacles and decorated with carved capitals and corner figures.

At the end of the sixteenth century, a new prison was built on a site immediately adjacent to the palace by La Paglia bridge – surely the most handsome place of incarceration. 'I think there is not a fairer prison in all Christendom', wrote Thomas Coryate in 1611. At the first-floor level of the palace it is linked by Antonio Contino's bridge which only in the nineteenth century became known as the Bridge of Sighs – the sigh being uttered by prisoners crossing from the courtrooms to the dungeons and getting one last glimpse of the Lagoon, the island of San Giorgio and the sun, sky and sea through its latticed windows.

This detail from 'Venice: The Molo with The Prisons and The Doges' Palace', by Canaletto, depicts the scene in 1743.

Thunder and rain! O dear, O dear!
    But see, a noble shelter here,
This grand arcade where our Venetian
    Has formed of Gothic and of Grecian
A combination strange, but striking,
    And singularly to my liking.

ARTHUR HUGH CLOUGH *Dipsychus*, Scene VII (c.1850)

Piazza Houses, Charleston
South Carolina, USA

19TH CENTURY

CHARLESTON was once the wealthiest and fourth largest city in the colonies, an east-coast seaport which thrived on trade along with New York, Boston and Baltimore. Its architecture reflected its affluence in timber, red or grey brick, and stone, and in a mixture of Classical styles. But the New World, as well as following European traditions a century or more afterwards, also developed its own response to the prevailing climate.

Pressure on available land saw the introduction of the terrace house with a single, gabled frontage to the street and the entrance offset to one side leading directly into a galleried porch running the full length of the house. Other common features of these 'side-yard' houses were doghouse dormers in the attic, Classical cornices, shuttered sash windows and a first-floor gallery often called the spinning gallery, where the women of the household spun cotton.

Behind the screen wall with its front door, the veranda was known as a piazza – so-called because of the open arcade which originally surrounded the first formal London square on the continental model, Covent Garden Piazza. The name was adopted as a general term for any open arcade, whether in wood or later (as in New Orleans) wrought iron. The piazza provided shade and captured any breeze to create a through draught – a welcome respite from the close atmosphere.

With its variety of fine domestic buildings, and the introduction of strict height ordinances for new buildings, Charleston trail-blazed the preservation of American cities more than 60 years ago. The oldest part of the town has been lovingly restored using a revolving fund so that the city which one anonymous critic described as 'Death on the Atlantic' had become, by the time Henry Miller wrote *The Air-conditioned Nightmare,* in 1945, 'a beautiful memory, a corpse whose lower limbs have been resuscitated'.

It more resembles a West Indian than an American city – from the number of wooden buildings painted white, the large verandas and porticos of the more stately mansions of brick, and the universal prevalence of broad verandas, green Venetian blinds, and other provisions to secure coolness and shade.

J S BUCKINGHAM *The Slave States of America in 1839* (1842)

The Red House
Bexleyheath, Kent, England

PHILIP WEBB 1859–60

For William Morris, the client and, later, leader of the Arts and Crafts Movement, the Red House was 'a small Palace of Art of my own'. For Philip Webb, his friend and architect, it was a first commission, designed when he was just 27. For Sir Nikolaus Pevsner, the architectural historian, it was the pioneer in the revival of English domestic architecture.

The simple, unpretentious house, built in an orchard in stark red brick (from which it took its name) challenged head-on the 'corruption' of nineteenth-century styles then prevalent, although it is also worth noting that it dates from the same decade as Joseph Paxton's revolutionary Crystal Palace. The architect took on the vestige of medieval master-builder once more, using indigenous materials and local craftsmanship to create an informal contemporary vernacular, more eclectic than pure Gothic in its inspiration.

Ruskin praised its truthfulness and hovering barn-like roof, 'which is its very soul … wherein consists its shelter'. Its windows dance to different tunes beneath the gables and ridges, hips and valleys. The L-shaped plan cradles within it the 'well-house', its high conical roof supported on massive oak posts and braces.

Tradition and innovation were combined in what looks more like a Victorian vicarage than a seminal building at first sight, a setting for weekend gatherings of like-minded artists who sought to counter the increasingly anti-social outcome of industrialisation.

The house was really a joint venture between client and architect, with Morris designing and executing much of the interior himself. He lived there for only five years; more recently, it has been owned and occupied by the architect Edward Hollamby for over 40 years.

It stands for a new epoch of new ideals and practices. Though the French strain which touched so much of the work of the Gothic revivalists is not absent, and the Gothic flavour itself is rather marked, every brick in it is a word in the history of modern architecture.

LAWRENCE WEAVER *Small Country Houses of Today*

Gatekeeper's Lodge
Park Güell, Barcelona, Spain

ANTONI GAUDÍ 1914

THE sign over the main entrance to Harrods department store in London reads 'Enter a different world'. That invitation could equally extend to almost any of the domestic work of Antoni Gaudí, as well as to his most famous building of all, the unfinished cathedral of Sagrada Familia.

His client for the Park, Eusebi Güell, was an Anglophile and admirer of the garden city which placed houses in a parkland setting (hence the English, rather than Catalan spelling of Park). It was to have had 60 houses built within it but in the end just two plots were sold, one to the architect himself and the other to a friend of his. The project turned into a financial disaster and, soon after the client died, it was sold to the city for use by the public.

The park is a fantastic world apart, with skewed columns, curvilinear staircases and park benches, and cave-like arcades. Sacheverell Sitwell described it as 'At once a fun fair, a petrified forest, and the great temple of Amun at Karnak, itself drunk, and reeling in an eccentric earthquake'.

But the care lavished by Gaudí on his earlier and grander houses and apartment blocks is also present in the two modest entrance pavilions which flank the wrought-iron gates, one a gatekeeper's lodge, the other an administration block. Exotic roofs of polychrome glazed tiles set in an icing layer of concrete provide a fifth elevation when viewed from the sloping park above them.

In 1984 UNESCO declared Park Güell a world-heritage site; eight years later, Gaudí's Casa Batlló, a residence he had extensively 'modernised', was bestowed with a different kind of honour altogether – inclusion in the *Guinness Book of Records* as the most expensive property in the world to be sold that year, at £50 million.

As you get to know Europe, slowly tasting the wines, cheeses and characters of the different countries, you begin to realise that the important determinant of any culture is after all the spirit of the place.

LAWRENCE DURRELL

Fallingwater, Bear Run Pennsylvania, USA

FRANK LLOYD WRIGHT 1936–9

FRANK Lloyd Wright's Fallingwater defies not only gravity but the structural engineers' prognosis that its long spans of reinforced concrete would collapse. In typically flamboyant manner, Wright buried their report in the living-room walls of rock, quarried from the site. The house also turned him into a star almost overnight. He is one of the few architects ever to grace the cover of *Time* magazine, in his case with Fallingwater as the backdrop.

The house, built for Edgar J. Kaufmann, appears to grow out of the rock in all directions, hovering over Bear Run waterfall. It was conceived as a series of horizontal 'trays' which were originally to be gilded in gold leaf, and vertical stacks of stone to simulate the stratification of the landscape. A huge natural rock was left in place to form the hearth of the fireplace which, as so often in the architect's houses, forms the hub of the spaces around it.

For Wright's 'rooms' were rarely enclosed by four walls; rather, areas were defined by a change in ceiling height or the position of built-in furniture, so that functions as well as spaces interpenetrated and overlapped, especially at the corners. What constituted the indoors and outdoors was deliberately left vague; even the continuous metal windows, made in England, touched the walls without a frame to emphasise the same natural materials used either side of them. The same stone floors seep out under the doors.

Wright submitted to no man, only to nature. From the relatively tight entrance hall, the house opens out to views on every side and invites the visitor to explore the cantilevered terraces where, at first, the rapid stream and its falls are heard and not seen; and then onwards down suspended stairs to the water itself. Wright's creative genius sings out, as in the words of Tennessee Williams: 'I don't want realism, I want magic'.

She has been looking at the external city; but the internal city is more important, the one that you construct inside your head. That is where the edifice of possibility grows, and grows without your knowledge: it is subject to no planner's control.

HILARY MANTEL *Eight Months on Ghazzah Street*

San Cristobál, Los Clubes Mexico City, Mexico

Luis Barragán 1967–8

Such is the ethereal quality of the work of some contemporary architects that it almost defies objective appraisal. The architecture of Luis Barragán is a case in point. Steen Eiler Rasmussen wrote that it was not enough to *see* architecture, that 'you must experience it', and in that context this Mexican engineer who taught himself the poetry of building readily springs to mind.

San Cristobál is a house, a stable for training thoroughbreds, and a series of pools – for horses and for people – designed for Folke Egerström and his family. Andrés Casillas was co-designer, but in its spirituality it is pure Barragán. Not surprisingly, Ricardo Legorreta and three generations of Mexican minimalists have become the followers of this architect whose inspirations included the fountains, pools and gardens of the Alhambra in Granada (which he visited in 1924) and the Guadalajara *pueblo* of his upbringing.

The daily ritual of bathing and exercising the horses is played out between the stables and a very long flat wall, punctured by two horizontal cut-outs which are high enough to walk or ride through. Beyond it, a glimpse of the surrounding landscape and gallops; to either side, a crude water-chute and a voluminous tree; a pond with a shallow sloping edge is the foreground – de Chirico's Surrealist world made flesh.

In Barragán's sensitive repertoire, his spaces – really outdoor rooms – become places, defined by the Dutch architect Aldo van Eyck as 'spaces with meaning'. Natural elements, in the form of planting, water and sky, are borrowed to complete the composition, set off by the vibrant colours painted onto the stucco walls.

Architecture is the manifestation of the dreams we desire.

RICHARD ENGLAND

Port-Grimaud
St Tropez, France

FRANÇOIS SPOERRY 1967–75 *(With later phases)*

FRANÇOIS Spoerry's luxury resort is in a very recent tradition of back-to-the-future residential developments, from the eccentricity of Sir Clough Williams-Ellis' Portmeirion, in North Wales, to Andres Duany and Elizabeth Plater-Zyberk's 'new urbanism' at the coastal resort of Seaside in Florida. Perhaps the first seeds were also sown here of Leon Krier's Poundbury development for the Duchy of Cornwall, on the edge of Dorchester (Thomas Hardy's Casterbridge), for the Prince of Wales visited and admired Port-Grimaud when serving in the Royal Navy in 1971.

It is a modern Venice, which Spoerry, both the architect and developer, visited in 1958. His vision was to clear away the mosquito-infested swamps between the old hill-town of Grimaud and smart St Tropez and recreate a Provençal-style fishing village in which every house would have two front doors – one to a street where traffic was carefully controlled, the other to a private mooring. Each house, from two to four storeys, is different within its pantiled pitched roof, washed walls in bright colours and individual features which enchant the eye.

Port-Grimaud is anathema, of course, to the purist, a pastiche rather than what its creator intended – an accumulated 'wisdom of the ages' which 'makes the heart want to sing'. Yet if imitation is the sincerest form of flattery, Spoerry is not without his flatterers across much of the Western world.

To those who live there, for perhaps six months of the year, it is an oasis; to sceptics, a wealthy ghetto devoid of social mix; to day tourists, a theme park with a market square, shops, cafés, hotels and artists' studios. It may be an 'instant' way of achieving an 'organic' result but its overriding qualities are its human scale, its variety and sense of identity – qualities absent from most 'instant' neighbourhoods of the past half-century.

Some cities have been planned geometrically while others have grown organically. In a modern grid-patterned metropolis there is little sense of local character or territorial identity. In the older, organically-grown city, every street corner is visually unique and the sense of precise location is much sharper, appealing more strongly to mankind's territorial sense.

DESMOND MORRIS *The Human Animal* (1994)

Samy House Dashure, Egypt

HASSAN FATHY 1979

I N spite of his rejection of many of the precepts of modern architecture when applied to his own country's needs – industrialisation of the building process, standardisation and prefabrication of building materials, the introduction of new forms and capital-intensive techniques, the artificial control of internal environments where traditional methods suffice (for example, air-conditioning versus using building mass) – Hassan Fathy has an extremely loyal following among architects in more developed countries than his native Egypt.

While it was inevitable that modernity and progress would be equated with these imported symbols, often in less than a generation, they were very often inappropriate in satisfying local require-ments. In a life devoted to housing the rural poor, but also to designing a small number of exquisite private villas, Fathy applied 'a mixture of social realism and Utopian vision'.

In 1940 he began experimenting with mud-brick structures, a tradition stretching back 5,000 years, and revived Nubian vault technology and the use of simple domes to span rectangular spaces. Although frustrated by official opposition and mind-numbing bureaucracy, he successfully reintroduced labour-intensive, self-build housing which produced better buildings and at lower cost. Time and resources were invested in local people whose standard of living and local economy benefited.

What was appropriate for the rural poor, particularly in his village of New Gourna, near Luxor, in the late 1940s, was also good enough for wealthier clients in terms of materials, forms and techniques. This weekend residence for Dr and Mrs Samy, built in limestone blocks and fired bricks, and finished in white plaster, displays no delusions of grandeur or pretence. The simple yet stunning carpentry of Hassan al-Naggar, who made the *mashrabiyya* window screens and the pergola which etches its intense shadow into an indoor/outdoor room, shows the dramatic effect that local craftsmen can still inject into architecture when given the opportunity.

Give a mason bricks and mortar, and tell him to cover a space and let in light, and the results are astounding. The mason, within his limitations, finds unending possibilities, there is variety and harmony; while the modern architect with all the materials and structural systems available to him produces monotony and dissonance, and that in great abundance.

JAMSHID KOOROS

French Ambassador's Residence Colombo, Sri Lanka

GEOFFREY BAWA 1979 *(Extension)*

ARCHITECTURE is Geoffrey Bawa's second profession but his first love. At the age of 32 he closed his law practice in Colombo and began to study at the Architectural Association in London. Returning home he soon began to receive commissions for houses, hotels, clubs, a university and the Parliament House in Kotte, all of which demonstrated where his true vocation lay.

Sri Lanka, formerly Ceylon, is known as 'the jewel in the Indian Ocean'. It achieved independence in 1948 after being a colony to successive outsiders – the Portuguese, the Dutch and the British. But for Bawa, all good buildings of these periods were first and foremost Sri Lankan, in response to the country's climate, its religious buildings and palaces, and its luscious vegetation.

The roof, steeply pitched and clad in the Sinhala tile, the half-round Mediterranean tile introduced to the country by Arab traders, is always the most important element because of the huge rainfall. The Dutch introduced the idea of raising the eaves of the roof even higher, to capture the breeze, and broaden them to shade walls and create verandas. The result, Bawa said, is a truly rational architecture which also offers the romance of beauty and pleasure.

His houses, including this extension, respond to both the immediate site and these traditions. Plans tend to be formal, with spaces and materials (brick, tiles, timber, plaster) clearly articulated around courtyards and passages where water plays a key role – both as pools and as rainwater cascading from gutter downpipes into huge bowls. The contrast of light and shade, different materials, solids and voids, and varying bright colours and textures, create a rich tapestry where inside, outside, and their interface merge imperceptibly.

If architecture is to be of service, it must be respond to more than need. The architect must also serve desire; the desire of the building to be what it wants to be and the desire of the human being for self-expression. In serving desire, architecture contributes to the spiritual enrichment of the world.

JOHN LOBELL *Between Silence and Light: Spirit in the Architecture of Louis I Kahn*

Krier House
Seaside, Florida, USA

Leon Krier 1989

Aｌｔｈｏｕｇｈ enormously influential as an urban theorist for more than 20 years, Leon Krier's first building was this three-storey Classical temple-cum-belvedere at the resort overlooking the Gulf of Mexico and started in 1981 by developer Robert Davis. Krier was a consultant on the whole project, masterplanned by Andres Duany and Elizabeth Plater-Zyberk.

The 'new urbanism' of the overall plan is anti-suburban: a place which has a focus in the 'public realm' or streets which link the privately-owned residences with an open-air market square, town hall, shops, post office and school. Civic buildings once again take on a special significance and all daily needs are catered for within a short walk rather than a car ride. The site area is 32 hectares (80 acres), about the size of a typical regional shopping mall in the USA, and the density 62 persons per hectare (25 persons per acre), higher than its suburban equivalent.

Seaside has helped to reassert traditional notions of 'community' and 'neighbourhood'. Key elements included a design charette, in which everyone who lives there was involved in a participation process with the consultants; and a simple building code (drawn up on a one-sheet poster) which regulates the form of what may be built by the different architects and self-builders. The net result has an harmonious feel even though there is a great variety of styles, from Victorian to Neo-Classical to Post-Modern to Neo-Modern – even Deconstructivist.

Like Port-Grimaud, Seaside has been criticised for catering for the wealthy 'snow-birds', those who can afford to relocate to warmer climes during the harsh northern winter, rather than providing a model for a balanced mix of social classes which is the intention at the Prince of Wales' Poundbury development at Dorchester, England, for which Krier is the masterplanner.

We shall not attain to cities and villages that are beautiful until we learn artistically to plan them. Transformations may help us greatly, as London and Paris and some examples at home show; but a mended article is never as good as one made well at first.

CHARLES MULFORD ROBINSON
The Improvement of Towns and Cities (1907)

The Jungle House
Tepotzlan, Mexico

SERGIO PUENTE AND ADA DEWES 1990–

THE extremes of nature and climate, so long considered hostile and alien to man's security and comfort and therefore something to be tamed, are instead welcomed into the Jungle House, situated near the Atongo river south-west of Mexico City. Designed as their own weekend retreat by Sergio Puente and Ada Dewes, building commenced four years ago and still continues on a steep hillside, amidst sub-tropical vegetation which grows so densely that photographer Richard Bryant had to hack a path through it to take his pictures.

The house – in fact two independent structures – responds in its siting to a spring at the northern and highest boundary, which feeds through the upper structure and into a plunge pool; and a meandering stream from a nearby waterfall around the lower structure. A long and dramatic flight of steps leading up to the latter, unguarded by handrails, recalls Aztec temples.

Adobe bricks were recycled from a house demolished nearby, and local people taught how to build walls incorporating pebbles from the river in the horizontal pointing. Roughly-cast concrete forms columns, beams, cantilevered floors and strongly articulated architraves.

Each of the structures is on two levels: the upper one using the spring water for washing and cooking, with a study – the only completely enclosed space – above it. The lower one has a living-room/bedroom on the ground floor, surrounded only by insect screens on three sides, with steps down to a shower and lavatory, and a roof-level kitchen and dining area above.

Few shelters today are so self-consciously open to the elements, exotic in their chosen location, and so inviting to those seeking to enjoy the luxuriant panoply of foliage which gently cradles the house in its arms.

Man's physical freedom manifests itself no doubt in his ability to choose the place on earth where he wants to live. Whereas immature reflection needs to judge by usefulness alone, a discriminating mind may ask its share of beauty. Neither privations nor danger will deter man from selecting a spot that provides him with the exhilaration generated by a superb landscape.

BERNARD RUDOFSKY
Architecture Without Architects (1964)

Ken Done House
Sydney, Australia

GLENN MURCUTT 1992

IN A series of houses and a few domestic-scaled buildings in rural locations, Glenn Murcutt has created the closest forms yet to an indigenous but contemporary Australian architecture. Earlier steel-and-glass pavilions, in the manner of Mies van der Rohe, later became iconoclastic extrusions of corrugated iron, with overhanging eaves, as either barrel vaults or triangulated pitched roofs. Corrugated iron has been readily adopted by Aborigines instead of traditional bark, due to their similar properties of flexibility, strength, lightness and sheet size.

More recently, however, and for suburban sites, other influences have come to the fore: the houses of Luis Barragán in Mexico and Antoine Predock in New Mexico, and Mediterranean vernacular architecture. Walls are often 'blind', to create privacy, and windows placed sparingly to give views out over 'chosen fragments' of scenery or activity. In the case of this house for the fabric designer and painter, Ken Done, and his wife Judy, the windows also give views into a courtyard which has a tree imported from New South Wales as its focus. Trees have a special place in his country as modifiers of climate and for Murcutt the roofs and louvres of his buildings are analogous.

This house in a wealthy suburb overlooking Sydney Bay is embedded into its steeply-sloping site and comprises two parts linked by a long, top-lit gallery along one boundary wall, used for hanging paintings. Three external 'rooms' form an integral part of it: a roof terrace, a grassed roof, and the sunken courtyard in between with its pool, patio and the tree. The plan is rational, the atmosphere Elysian. The long barrel vault, which he once likened to the arch of the sky, is still present, but as a ceiling to the living-room and kitchen; and shutters, louvres and blinds all emphasise the horizontal at the expense of the vertical in perhaps an unconscious reference to Australia's unending horizons.

The place to improve the world is first in one's heart and hands, and then work outward from there. Other people want to talk about how to expand the destiny of mankind. I just want to talk about how to fix a motorcycle. I think that what I have to say has more lasting value.

ROBERT M PIRSIG *Zen and the Art of Motorcycle Maintenance* (1974)

Assembly House Siófok, Hungary

Imre Makovecz 1988

Bela Bartók, the Hungarian composer, spoke for all 'contemporary regionalists' in architecture when he said 'What is new and significant must be grafted onto old roots'. Imre Makovecz and his practice, Makona, invest in their architecture a love of local culture and enduring traditions which is shared by no more than a dozen near contemporaries in different countries: Hassan Fathy in Egypt, Luis Barragán in Mexico, Rifat Chadirji in Iraq, Richard England in Malta, Geoffrey Bawa in Sri Lanka and Glenn Murcutt in Australia among them.

Makovecz is a maverick who suffered internal exile for 30 years designing modest buildings, before he lept onto the world stage with the Hungarian Pavilion at Expo 92 in Seville. Beneath the upturned hull of timber and slate was a large oak tree from the Gemenc forest, the Tree of Life, planted in a glass floor so that its roots were as visible as its branches. It was a symbolic statement about time past, present and future, and the need for an organic architecture rooted in spiritual and cultural values. He talks about the angels in Wim Wenders' film, *Wings of Desire,* coming to earth and being made flesh. For him the angels become architecture.

His approach here at Siófok, near Lake Balaton, which comprises the home of a Lutheran priest and his family, guest rooms and an assembly hall, combines his pre-industrial vision with a craft approach, borrowing and adapting native forms, building materials and techniques to create something which is at once reassuring and surprising.

Makovecz has a passion for the works of Frank Lloyd Wright and the anthropomorphism of Rudolf Steiner's philosophy, and yet each of his buildings responds directly to its site and the client's brief – the client being taken in the broad sense of those who will use the building as well as the individual who commissions it. With the next generation of his followers, Makovecz is helping to redefine Hungary's role as a nation rather than as a satellite of either East or West.

'Where is the plan you are following, the blueprint?'
'We will show it to you as soon as the working day is over; we cannot interrupt our work now', they answer.
Work stops at sunset. Darkness falls over the building site. The sky is filled with stars.
'There is the blueprint', they say.

ITALO CALVINO *Invisible Cities* (1974)

The Tent House, Eumundi Queensland, Australia

GABRIEL POOLE 1992

For D. H. Lawrence, Australia was 'a soft, blue, humanless sky' with a 'pale, white unwritten atmosphere', a vast, expansive *tabula rasa* 'without a mark, without a record'. In his novel *Kangaroo,* of 1922, the landscape was depicted as a bleak and lonely place: 'It is said that man is the chief environment of man. That for Richard, was not true of Australia. Man was there, but unnoticeable.'

When the architect Gabriel Poole and his artist wife Elizabeth Frith decided to build a new house for themselves on the outskirts of Eumundi, a small town famed for its beer 60 miles north of Brisbane, they wanted something that would harmonise with rather than dominate the surroundings by treating them as hostile. The Aborigines have a saying for this approach – 'to touch-the-earth-lightly' – and the steel-framed, canvas-walled Tent House does just that.

Designed to cope with the extremes of climate – heat, humidity, tropical rainstorms and the occasional cyclone – draughts of air are caught by louvres over the balcony and channelled through the rooms. Walls can be rolled up so that butterflies and giant cicadas fly through, oblivious to their trespass. The artist's studio, a separate structure nearby, doubles as a spare bedroom with only two walls and a roof for shelter. Nature provides not only air-conditioning but also solar radiation for power and hot water. The rain forest on their 20-hectare (50-acre) estate is being cultivated once more around the Bunya pines, revered by the Aborigines for their mystical powers.

Similar houses are being developed by Poole in kit form and were first marketed at Brisbane Botanic Gardens in 1992. They can be assembled in single-storey, two-storey and split-level versions, as large or small as determined by space requirements and budgets. A house like the one illustrated costs little over £40,000 to build.

In dwelling, be close to the land.
    In meditation, go deep in the heart.
In dealing with others, be gentle and kind.

LAO TSU *Tao Te Ching* (Sixth Century BC)

Stone and Brick Towers
Sana'a, North Yemen

*c.*2,000 BC

LEGEND has it that Shem, Noah's son from whom the Israelites were descended, founded the prehistoric stronghold of Ghumdan upon which Sana'a now stands. The city dominated trade 4,000 years ago due to its strategic location at the southern entrance to the Red Sea, and records show that at that time stone houses of four to six storeys were built. The castle, which dates back to before AD 100, had 20 storeys and the city walls (Sana'a means fortified town) were six to nine metres (20 to 30 feet) thick.

Few Europeans visited the place until the middle of the twentieth century and nature played its part in keeping out the unwelcome: the city is situated in the highlands, 2,350 metres (7,700 feet) above sea level. Stone, and later brick, succeeded in overcoming the threat of one of the region's most subversive and destructive invaders, termites. There is a strict hierarchy of uses on different floors.

The Arabs were the first to make bricks, from clay available on river banks. Packed clay gave way to prefabricated and rectangular mud and straw blocks formed in wooden frames. After evaporation, the frames were removed and the bricks left to dry in the sun. Later, about 3,000 BC, in Mesopotamia, it was discovered that bricks baked in kilns were stronger and these gradually replaced their sun-dried predecessors.

Most of the 14,000 towers still standing today are six or more storeys high and date back more than 400 years. They have withstood the ultimate test, time. Mud was used to create flat roofs over timber beams. Specially shaped bricks produced more and more intricately decorated openings for doors and windows, which were then coated in a thick layer of gypsum and painted with a limestone whitewash. Brick specials were often used in conjunction with geometric tile patterns in some parts of the Middle East and, as basic building materials became more modular and prefabricated, so too shelters of all types became more rectangular or square in plan and elevation.

Come let us make bricks and bake them hard. Come, let us build ourselves a city and a tower with its top in the heavens, and make a name for ourselves; or we shall be dispersed all over the world.

GENESIS chapter 11, verses 3–5

Caernarvon Castle Gwynned, Wales

MASTER JAMES OF ST GEORGE 1283–1323

Sir Banister Fletcher wrote: 'The thought that went into castle-building was of a different order from that which produced cathedrals – intelligence rated higher than imagination. But as images of power they would be every bit as eloquent as great churches.' Edward I's castles were in this sense more impressive than anything else of the period outside the Holy Land.

Master James of St George, the royal mason, accompanied the king on his Crusade. It is likely that he visited the fifth-century Theodosius at Constantinople on which the design of Caernarvon is thought to be based. Its irregular shape and banded masonry walls are testament to that. To the rebellious Welsh, the castle must have been a forbidding place on a quite extraordinary scale. At 38 metres (124 feet) high, the Eagle Tower is one of the tallest structures of the Middle Ages.

The castle was intended to fulfil a greater purpose than subjugation, however. It was built as a royal household and administrative centre. The king had the prescience not only to ensure that his first son and heir was born within its battlements, but that this first Welsh prince should be invested there as the first Prince of Wales as well. Edward of Caernarvon, later Edward II, was also the last heir to the throne to be born in Wales. The most recent investiture at the castle was of the present Prince of Wales in 1969.

Surrounded on three sides by water, Caernarvon Castle is a more sophisticated version of the Norman motte-and-bailey. Its single encircling wall was both stronger and higher than any of its antecedents and has 13 polygonal towers around its perimeter. Masons and carpenters were brought in from throughout England and the castle's estimated cost was more than £40 million at today's prices. Even so, it was never formally completed, as the Scots proved to be even more troublesome than the Welsh.

> Though I have kind invitations enough to visit America, I could not even for a couple of months, live in a country so miserable as to possess no castles.
>
> JOHN RUSKIN (1889)

Towers of San Gimignano Tuscany, Italy

12TH AND 13TH CENTURIES

Within the walls and gates of San Gimignano (population 7,700) rise the 13 remaining stone towers of the best preserved medieval hill-town in Tuscany, just 10 miles north-west of Siena. Built largely in the twelfth and thirteenth centuries, the towers were thought at one time to number 76. When seen from a distance they gave the impression of an 'urban pin-cushion', observed Lewis Mumford. But many of them were too ambitious for their foundations and collapsed, so that by 1580 only 25 were left.

Built as adjuncts to the palaces of the town's powerful and rival families, they must have proved a challenging climb even for the fit, 600 years before Elisha Graves Otis patented his safety lift. The Torre della Rognosa, next to the Palazzo del Podestà, rose 51 metres (167 feet) and prompted a statute to discourage even more vainglorious plans. Yet within a few decades work was begun on the Torre Grossa, three metres (10 feet) higher.

The towers of San Gimignano were built for ostentation and prestige rather than for defence or as look-outs; and wherever they are found today, the tallest buildings continue to symbolise power and supremacy.

In terms of competing families and height, even San Gimignano's skyscrapers are over-shadowed by those at Bologna. That city once boasted 200 towers and the Torre Asinelli, built in 1109, still stands today at 69 metres (226 feet) – the equivalent of a modern 22-storey tower block.

Such one-upmanship is still commonplace between cities, New York and Chicago for example, or within the same city. Most famously, the proposed height for New York's Chrysler building in the 1920s was 282 metres (925 feet); but when a rival bank announced it was to top it by less than one metre, Chrysler's architect secretly assembled a spire to add to his skyscraper so that it ended up at 319 metres (1,046 feet) and set a new record for the then tallest building in the world.

Siena sat enthroned upon her hills in knightly vigil, and so close she seemed in that spun air that I might have shot an arrow into her, or perhaps have tossed a gauntlet into the Campo, or maybe have shouted some taunt that would have been clearly heard and resented. Such thoughts are natural in the shadow of these towers.

H V MORTON *A Traveller in Italy* (1964)

Matsumoto Castle
Honshu, Japan

16TH CENTURY

BERNARD Rudofsky, in *The Prodigious Builders* (1977), wrote: 'Like Japanese armour, Japanese castles have a flavour all their own. The cyclopean walls that rise from the moat, of an elegance unknown in the Western world, form a strange contrast to the highly flammable wooden structures inside.' But what structures they are: secular pagodas spreading their ornate gables and eaves into the sky, the more lofty and elaborate the more powerful the shoguns, or hereditary feudal lords, who owned them. But they also had a practical purpose: trapdoors underneath the eaves allowed the occupants to shower missiles down on attackers.

Matsumoto Castle, along with Kumamoto and the 'White Heron' of Himeji, was the first in a new breed of fortresses built on the flat ground of the lowlands rather than in elevated positions. For the first time the introduction of firearms became a major design consideration, so that walls were thickened, moats deepened, towers heightened and masonry was used to reinforce the timber-shuttered gallery walls. Timber was always the major Japanese building material, but the sloping bastions were made from crudely shaped stones, sometimes (as at Osaka Castle) as big as a four-storey house.

While we relish the castles for their architecture, until a century ago the Japanese were embarrassed by their medieval associations. Many were destroyed, sold off in job lots for, among other things, firewood, until an active programme of restoration and reconstruction was commenced in the 1950s. Such has been their pulling power in recent years that, as Rudofsky wrote, 'Today, Japanese castles are built as local attractions, and by no means for foreigners only, in places where there had never been castles before'. The genuine examples are now protected as national treasures.

And what is it you guard with fastened doors?
Have you peace, the quiet urge that reveals your
    power?
Have you remembrances, the glimmering arches
    that span the summits of the mind?
Have you beauty, that leads the heart from things
    fashioned of wood and stone to the holy
    mountain?

KAHLIL GIBRAN *The Prophet* (1926)

Neuschwanstein Castle Bavaria, Germany

Eduard Riedel and Georg von Dollmann 1869–81

MAD King Ludwig II of Bavaria had a passion for building more and more extravagant structures. As well as Neuschwanstein he commissioned a replica of Louis XIV's Versailles and another based on the Trianon. Several were never completed.

Precariously perched on top of a rocky crag overlooking the Alps and its lowlands, the castle is a shrine to Richard Wagner. The king, who ascended the throne in 1894 at the age of 18, immediately formed a mutual appreciation society with the composer, paid off his debts and became his lifelong patron and promoter. Wagner's operatic works inspired the design and it is more of a stage set than a fortification ever intended to repel attack.

The German legends which fuelled Wagner's creativity are reflected in many of the rooms. The Singers' Hall echoes *Tannhauser,* the opera set in Wartburg, and *Parsifal,* the sacred festival drama and Wagner's last work. Below, in the king's private apartments, the theme is *Tristan and Isolde.* The Throne Room, by contrast, is Byzantine. But the castle did not lack for creature comforts. An early type of central heating was installed and there was hot and cold running water in the main kitchen.

Declared insane and replaced by his uncle as regent in 1886, Mad Ludwig did not have long to enjoy his fantastic creation. He drowned himself, and his psychiatrist also perished while trying to save him. On a happier note, through Disneyland's romantic Cinderella Castle – the theme park's trademark – the king's aspirations live on, although the American interpretation is more French than Bavarian in its styling.

If a man is created, as the legends say, in the image of the gods, his buildings are done in the image of his own mind and institutions.

LEWIS MUMFORD

Highpoint I, Highgate Hill London, England

BERTHOLD LUBETKIN AND TECTON 1933–5

IT IS hard to imagine now the excitement there was among a new generation of British architects when the first of two neighbouring tower blocks was built at the summit of Highgate Hill half a century ago. The architect was a Russian émigré in his early thirties, Berthold Lubetkin, who surrounded himself with other like-minded young architects including Denys (later Sir Denys) Lasdun, who went on to design the National Theatre on London's South Bank. For these flats, and the famous penguin pool at London Zoo in Regent's Park (completed 1934), they teamed up with a brilliant young English-born Danish engineer, Ove Arup.

The design of the eight-storey cruciform block was heavily influenced by the ideas of Le Corbusier who, in 1926, published his *Five Points of a New Architecture,* a reference to the five classical orders of architecture. Among these were the raising of the building above pilotis (columns) in order to free up the ground-floor space; a garden terrace on a flat roof; and a preference for ribbon windows. Highpoint scored three out of five. Its load-bearing reinforced concrete walls did not permit flexible interior planning or the liberation of the external walls from rigid geometry, however.

Le Corbusier and his myriad disciples also approved of its finish in white render and the overall detachment of the building from its surroundings. It was, he said, 'one of the first vertical garden cities of the future' – high praise indeed. It would have been even better had it been social/Socialist housing for the masses; instead it provided luxury apartments and maids' quarters.

When Highpoint II was completed next door, three years later and after local opposition, the architectural profession was shocked to see classical caryatids (carved female figures functioning as columns) holding up the entrance porch, which they considered a betrayal of Modernist principles and a harking back to a past they had largely rejected in design.

> The garden city is a will-o-the-wisp. Nature melts under the invasion of roads and houses and the promised seclusion becomes a crowded settlement ... The solution will be found in the 'vertical garden city'.
>
> LE CORBUSIER

Unité d'Habitation
Marseilles, Provence, France

Le Corbusier 1947–52

I ts architect exclaimed, 'This is the building I have wanted to build for 25 years', when finally given the opportunity to practise what he had preached. Utopia, according to Le Corbusier, comprised a forest of tower blocks set in open parkland, a 'vertical city … bathed in light and air'. Unité at Marseilles was the first post-Second World War megastructure, a self-contained village of 1,600 people in 337 apartments of 27 different types, with internal 'streets' for shopping and communal facilities half way up, and a nursery, gymnasium and running track 19 storeys above ground on the roof.

The colossus, analogous to an ocean liner, measures 165 metres (540 feet) long, 24 metres (79 feet) wide and 56 metres (172 feet) high – 'a temple of family life', an antidote to suburban sprawl.

Originally conceived as a steel structure, it was built in rough concrete due to materials shortages. Paint does little to relieve the greyness. The pilotis (columns) on which the building sits, are two-storey-high dinosaur legs, even though Le Corbusier had written, in less politically-correct times, that they should be like 'the strong curvaceous thighs of a woman'. Each apartment is slotted into the skeleton grid, like 'bottles in a wine rack'; most have double-height living rooms, deep balconies and even deeper floor plans, restricting daylight at the centre.

This, and subsequent Unités, proved enormously influential, for example, at the Roehampton Estate (1952–5) on the edge of Richmond Park, London, although without all the social facilities. Soon local authorities throughout Britain, and elsewhere, tried to outdo each other in the number and height of their clones, often with disastrous consequences. Le Corbusier's rational, radical vision was ultimately stultifying when applied *ad infinitum, ad nauseam* by lesser mortals. His least-quoted phrase, which arose when challenged over the inhabitants' modifications to another housing scheme, is: 'It is always life that is right and the architect who is wrong'. The French government listed Unité as an historic monument in 1964.

If we eliminate from our hearts and minds all dead concepts in regard to the houses and look at the question from a critical and objective point of view, we shall arrive at the 'House-Machine', the mass-production house, healthy (and morally so too) and beautiful.

LE CORBUSIER *Towards a New Architecture* (1923)

The Hundertwasser-Haus
Vienna, Austria

Friedensreich Hundertwasser and Peter Pelikan 1983–5

VIENNA boasts more than its fair share of idiosyncratic architecture, largely as a result of spawning, or educating, or commissioning, a succession of maverick architects – Otto Wagner, Josef Hoffmann, J. M. Olbrich, Hans Hollein and Günther Domenig readily spring to mind. Its programme of social architecture dates back to the 1920s; but celebrating the art of architecture has always been a major concern.

The Hundertwasser-Haus is a landmark building in the city's Third District, commissioned by the city fathers and a collaborative effort between the artist Friedensreich Hundertwasser and the architect Peter Pelikan. It is high-rise housing with a human face, 'a defiant gauntlet thrown in the face of the monotonous boredom of rational architecture', a 'poetic antidote to anonymous, hostile, aggressive deserts of concrete'.

The materials are brick, timber, plaster and tile – an elaborate fusion of surprising contrasts and contradictions. The building has a mixture of uses, with shops, a doctor's surgery, terrace café and multi-purpose rooms in addition to 50 apartments. And then there is the roof: 150 trees, grass and flowers planted in 500 tons of soil, amidst bulbous onion domes reminiscent of the shimmering paintings of the Art Nouveau artist Gustav Klimt.

It has been variously described as 'eco-kitsch' and 'a successful transition to a more ecologically-oriented urban culture'. Soon after it opened, the city issued an open invitation for people to come and take a look; 760,000 responded, taking in the Roman fountain at the entrance, walking over the undulating floors of the hallway which are intended as a 'melody for the feet', and observing more than 50 statues and other artworks. Now it is on every tourist's itinerary.

162

When the Stranger says: 'What is the meaning
    of this City?
Do you huddle close together because you
    love each other?'
What will you answer? 'We all dwell together
To make money from each other?' or 'This is a
    community?'

T S ELIOT *The Rock*

The Walled City Kowloon, Hong Kong

20TH CENTURY

ALMOST 2,000 years ago Seneca was moved to write to his friend Lucilius: 'Life is the gift of the immortal gods, living well is the gift of philosophy. Was it philosophy that erected all the towering tenements, so dangerous to the persons who dwell in them? Believe me, that was a happy age, before the days of architects, before the days of builders.'

What would he make of the Walled City, alias the 'City of Darkness', Hak Nam, the 'Anarchist's Delight'? Here an estimated 33,000 people live in one great densely-packed, organic mega-structure rising up to 14 storeys on an area of just two-and-a-half hectares (six acres) – the world's most densely populated place of shelter. But in this case it was not architects who erected the 'towering tenements', nor professional builders, but the inhabitants themselves. Its effect is at once shocking, exhilarating and foreboding, a late twentieth-century inferno of crime and squalor.

During the last half-century the place has grown without regulation or restraint due to an anomaly in Britain's terms with China for the 99-year lease on the colony negotiated in 1898. It is in no-man's land – within the colony's domain, yet outside its control. The Chinese, virtually all refugees, live among themselves, 'untaxed, uncounted, untormented by governments of any kind', wrote Peter Popham in the *Architectural Review* in 1993. 'The Walled City became that rarest of things, a working model of the anarchist society.'

Its people live above, below and beside the shops, the textile factories, the gambling and opium dens and the brothels. There are also schools and a church but no streets, only alleyways often little more than a metre (three feet) wide, with bicycles the only means of transport within it.

As a human habitat it is unique, but perhaps not for much longer; many believe it has been tolerated for far too long already as 'the closest thing to a truly self-regulating, self-sufficient, self-determining modern city that has ever been built'.

There, sighs, laments and loud wailings resounded through the starless air ... Strange tongues, horrible cries, groans of pain, cries of anger, shrill and hoarse voices, and the sound of beatings, made a tumult, circling in the eternal darkness, like sand eddying in a whirlwind.

DANTE *Inferno*, iii, 22–30

Millennium Tower
Daiba Bay, Tokyo, Japan

SIR NORMAN FOSTER AND PARTNERS 1989 *(Design commissioned)*

SKYSCRAPERS dominate twentieth-century cities in a way that the great cathedral-builders of the Middle Ages could never have imagined possible. To new materials and techniques, such as iron and steel and reinforced concrete in skeletal construction, have been added the safety lift, electric light, air conditioning and telecommunications. The rest is a recent history of record-breaking structures, mostly in the New World, and dominated by New York and Chicago. Now their hegemony is being challenged by countries of the Pacific Rim, notably Japan.

Sears Tower, Chicago, completed in 1974, is currently the tallest building in the world at 442 metres (1,450 feet). If Sir Norman Foster's offshore Millennium Tower for Tokyo is ever built, it will be almost double that at 840 metres (2,756 feet). And it will fulfil Chicago architect Louis Sullivan's prophecy of a century ago that a tall building 'must be a proud and soaring thing, rising in sheer exultation from top to bottom without a single dissenting line'.

The one million square metre (10,764,000 square feet) helical spike is designed to shelter 50,000 people. It is, in effect, a vertical town in its own right, with exclusive residential apartments near the top, sandwiched between public observation platforms, restaurants and bars at the summit, and offices, shops and leisure facilities below – a diversity which deliberately contradicts modern town planning's preoccupation with zoning different human activities. Technology already exists to realise its construction; whether any financial institutions consider it viable and a worthwhile risk is still being tested.

A millennium is meant to be a period of good government, great happiness, and prosperity. The early hominids, who first built protective structures almost 400,000 years ago, could have had no concept of such aspirations. Just how far we have evolved in meeting them is a moot point on which to end this brief survey of shelters, from around the world and throughout time.

Make no little plans. They have no magic to stir men's blood and probably themselves will not be realized. Make big plans; aim high in hope and work ... Let your watchword be order and your beacon beauty.

DANIEL BURNHAM (1907)

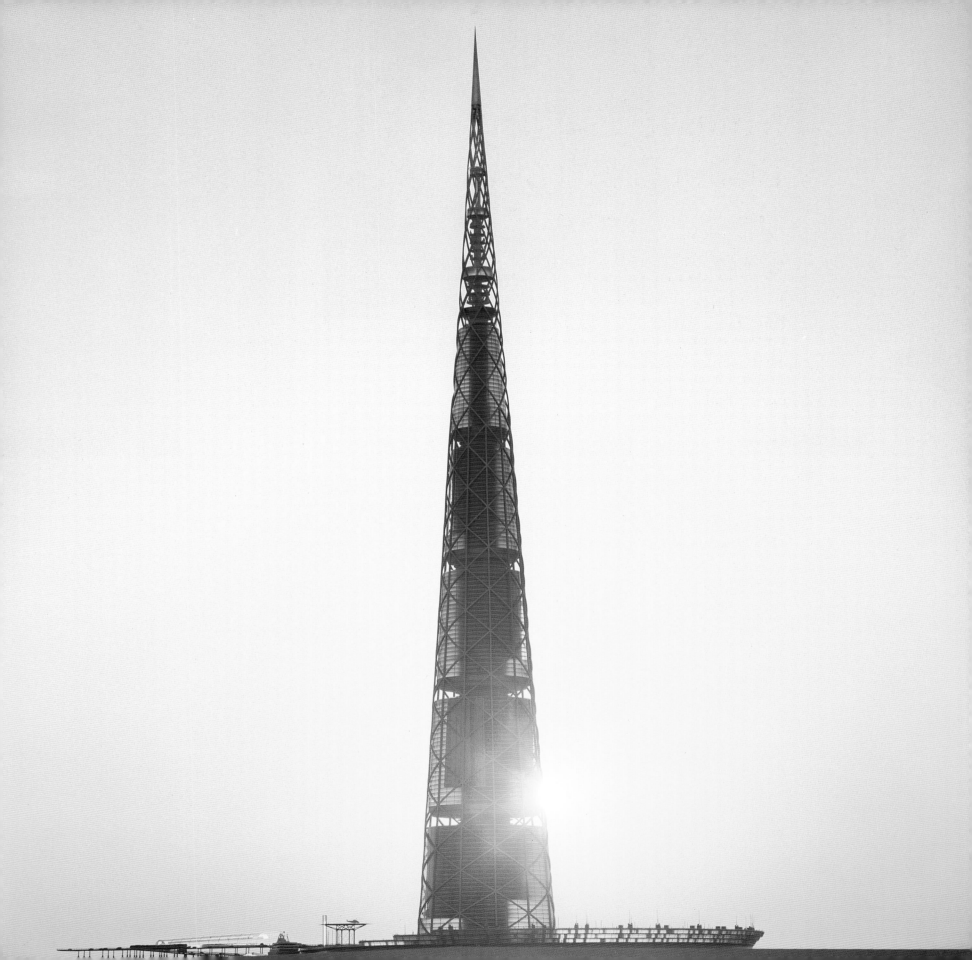

INDEX